John Heard

**Breeding, Training, Management And Diseases of the Horse And**

**Other Domestic Animals**

John Heard

**Breeding, Training, Management And Diseases of the Horse And Other Domestic Animals**

ISBN/EAN: 9783744724623

Printed in Europe, USA, Canada, Australia, Japan

Cover: Foto ©berggeist007 / pixelio.de

More available books at **www.hansebooks.com**

# BREEDING,

# Training, Management

## AND

# DISEASES

### OF THE

# HORSE

### AND OTHER

# DOMESTIC ANIMALS,

*With Ninety-Five Illustrations.*

—BY—

## J. M. HEARD, M.R.C.V.S.

*Late Professor of Veterinary Surgery in the New York College of Veterinary Surgeons ; Member of the New York Academy of Comparative Medicine and Surgery.*

Author of "Horse Shoeing, Past and Present," "How to tell the Age of a Horse," " Chart for Horse Owners," etc.

**PUBLISHED BY THE AUTHOR.**

NEW YORK, 1893.

# PREFACE.

ALTHOUGH there is no dearth of books which treat on the subjects contained in this little volume they are, for the most part, so expensive as to be out of the reach of the average American farmer, who must, after all, when considered in the aggregate, do the great bulk of breeding, and they must necessarily be the owners of a vast majority of the domestic animals contained in the whole country.

Again ; they treat, for the most part, on special subjects and, therefore, contain a great deal of matter that is useless and not clearly understood by the average stockowner. Neither has he the time to devote to a complete and detailed study of each of the subjects treated in this work. It will also be found that most books on Breeding and ▪▪▪▪ ▪▪▪▪ ▪ ▪▪imals that are

.

written in plain, every-day language are very ancient and, therefore, have not the information contained in modern books, which may be based on the vast advance which has been made in scientific breeding, and on the great discoveries which have been recently made in medicine and surgery.

It has been my aim to give the average stockowner the information that he must necessarily be continually seeking, in language so plain, that any common school-boy can understand it. I have, therefore, avoided technical terms wherever possible.

Another reason which has impelled me to arrarge a book relating to diseases of animals, is the fact that in a large extent of our country, the services of an educated Veterinarian are difficult to obtain by the average stockowner who resides, for the most part, at a considerable distance from large towns, and even when obtainable, it will be only at a considerable expense, besides the loss of valuable time in checking serious disease and thereby avoiding a part of the losses which in the aggregate, are enormous. A recent bulletin issued by the Department of Agriculture, placing the annual loss of live stock in the United States by disease, at more than $100,000,000. From experience, I am sure that a large part of this immense loss can be avoided, and it has been my aim to give the necessary advice which will eventually lead to the saving of a very large percentage of the domestic animals which at present die from preventable or curable diseases.

The insertion of the section relating to the administration of medicines, with an alphabetical table of diseases and their remedies and the table relating to medicines and their doses was suggested to me by Dr. J. A. Breakall, for which I thank him. My brother, Dr. A. M. Heard, has also rendered valuable aid in many ways which I cannot here specify. I also thank Mr. George Kittredge for the interest he has taken in making the drawings for several of the illustrations, and Messrs. Clarke and Richardson, for the loan of valuable cuts.

**J. M. HEARD, M. R. C. V. S.**

**NEW YORK CITY.**

# Copies of this Book

may be ordered through any

# PUBLISHER, BOOKSELLER

or

# NEWSDEALER

in the

## UNITED STATES AND CANADA

or will be sent free by mail on receipt of price

(One Dollar)

by

DR. J. M. HEARD, M.R.C.V.S.

119 WEST 56TH STREET,

NEW YORK CITY, N. Y.

# PART I.  BREEDING.

## INTRODUCTION.

I AM well aware that there is no royal road that will
lead to the highest degree of success for all who may
attempt the task of becoming profitable breeders of
high class animals, whether of horses, cattle, sheep,
swine, poultry, dogs, or in fact any of the domesticated
animals. In fact it may be generally stated that the
failures will outnumber the successes as ten to one. On
looking over the field, however, we can see many man-
agers who are successful, and on careful inquiry it will
be found that nearly all of them have attained success
because they have acted on lines which are entirely in
accord with proven biological facts. It may be asserted that success in breeding will not be achieved by
any hap-hazard method, but by careful plans, laid out
after great deliberation, and according to a certain
standard of action which breeders have set up for their
guidance. I am aware that an occasional hap-hazard is
sometimes temporarily successful, and this may occur
apparently in spite of the violation of what many successful men would call positively correct principles.
These facts show to some extent the great difficulties
with which the breeder has to contend; especially will
the difficulty appear very great if we bear in mind the

almost fabulous amounts that are yearly spent by some of our wealthy men in efforts to become successful breeders of one or more species of domestic animals. On close scrutiny it will be found that there are certain well defined principles which it is necessary for all breeders to adopt if they wish to be permanently successful, especially in a pecuniary sense. And perhaps it may not be out of place here to sound a word or two of warning against what seems to be a tendency to the production of a degenerate race, so far as horses are concerned.

I presume it will be admitted that the aim of select breeding in the domestic animals should be to obtain an animal that is most suitable for the accomplishment of the labor that is required to be performed by that species. Of course it is too much to ask breeders to forego a prospective present profit for the good of the future generation of horses, for there are very few men with public spirit enough to refuse to breed horses that can run a fast half mile, and for which they can obtain an exceedingly profitable price, notwithstanding that in most cases at the present time those same horses are entirely useless for any other purpose, not even making good hacks or hunters, and certainly useless on a farm. And the numerous small race tracks in the neighborhood of large cities where short races are being run all the year round seem to increase the evil to a tremendous extent, and will probably lead to the development of immense speed for short distances, to the detriment of the general advancement of all breeds of horses that would naturally take place were speed and stamina combined to be the qualities necessary to profitable ownership. After all, it is perhaps easy to exaggerate the presumed ill effects of such periods of fashionable aberrations as above described, for nearly all of the vast numbers of animals that are annually bred in all civilized countries must necessarily be raised for practical purposes. This will always be the underlying basis for the general breeder, no matter how some breeders may succeed in specializing a certain number of animals, and this assertion holds good in the case of all domestic animals. A horse that may be considered a typical animal for general purposes is seen in Fig. 3.

Fig. 3. Hackney.

## GENERAL PRINCIPLES.

THERE are certain biological laws which relate to Re-
production, to which all animals are subject. Many of
those laws have only been formulated within the past
half century; in fact, to the influence of Charles Darwin
more than any other man, or perhaps any ten men, is
due the formulating of reproductive laws and the
grouping of facts relating to them.

The two underlying principles each of which is opposed to the other and by which all breeders are
guided, are Heredity and Variation.

## HEREDITY.

WHEREVER we look in the field of Natural History,
the one fact that stands out more prominent than any
other is the strong tendency for like to beget like; yet
if this were an invariable rule there could be no change
in types of animals, and there could be no advancement in breeding. Again, unless the principle of
heredity was exceedingly strong, there would be no
permanent fixity of any improvement which might be
produced by any scientific breeder. But under the
strong influence of the law of heredity any chance
improvement which may occur in any breed of animals,
can almost certainly be reproduced in future individuals that may be born as a result of pairing the animal
in which the improvement has appeared, with another
of the same species. The failure to reproduce this im-

provement in the progeny will frequently happen, but by suitable pairing the improvement can generally be reproduced. If we can lay down a rule to guide us in accomplishing this very desirable object, a great advance will be made in scientific breeding. This we shall now attempt.

It has been found that in some families of animals a tendency to reproduce their like is much greater than in others. I will only mention one which will be familiar to breeders all over the country, namely, "Rysdyk's Hambletonian." It used to be a very common observation: "How much like the old horse he or she is," even to grandsons and grand-daughters of this horse, in form and style as well as in action. This horse had a great individual tendency to reproduce this improved form, style and speed, even when mated with very inferior mares. This tendency in one parent to overcome what is lacking in the other is called "prepotency." We say therefore that in Hambletonian his prepotency to produce his desirable qualities was very great. It may also be said that the prepotency of Electioneer was also very great, in fact, so great was it that it seemed to make no difference scarcely what kind of mare he was mated to, a fast colt was sure to be the result. Even to thoroughbreds his influence was so much greater that fast trotters were almost invariably the result of the pairing, a result of such a mixed union being the celebrated Palo Alto 2.08¾—Electioneer himself, being a son of Rysdyk's Hambletonian, thus inheriting a very strong tendency to reproduce fast and well formed trotters, as we have seen. How are we to tell with reasonable certainty, whether an animal is endowed with unusual prepotency or not? If it were possible to tell with certainty, we could almost class breeding with the exact sciences. Unfortunately we cannot do this except in those cases where the prepotency has been proven, as was the case with Hambletonian, Electioneer, and others after a trial at the stud for three or four years. But the following rule will be found useful; *viz.* : that the longer the line of descent in which the desired trait can be traced, the more likely is the desirable quality to be transmitted to offspring; so that any quality which may arise in an individual, if transmitted for several generations, will become stronger and stronger with each generation, and will soon become strongly inherited, and the prepotency of each descending parent will be increased in the direction of the altered character. Hence, all other things being equal, there would be much better chance for obtaining a very fast thoroughbred, with lots of stamina, by mating a female with a male that had proven fast for long distances; and this chance would be greatly strengthened if the male progenitors had also inherited great speed with good staying powers. Therefore, the greater the number of progenitors in which can be traced the desired quality, the more likely is the quality to be reproduced in the offspring. This rule applies to all animals, whether wild or domestic.

Unfortunately we have a prepotency to reproduce undesirable qualities as well as desirable. This is one of the great drawbacks that breeders have to contend with. For instance: a sire with small, thin feet—especially if it has been inherited for two or three generations—will be likely to reproduce in a large proportion of his progeny this very undesirable quality. There are many other imperfections and many diseases and malformations which are inherited. The following are some of the diseases which are strongly inherited:

Anchylosis of Joints (Stiff Joints)

Broken Wind.

Cataract.

Curb.

Diarrhea.

Hemorrhage.

Imperfections of Feet.

Navicular Caries.

Ophthalmia. (Moon Blindness).

Ostitis or Sore Shins.

Rheumatism.

Rickets.

Ringbones.

Roaring.

Sidebones.

Spavins.

Staggers.

**ANCHYLOSIS of JOINTS.**—As a result of various inflammatory diseases in bones and joints we very frequently get a complete union between the two or more bones that enter into the formation of a joint. There are some joints in horses that are peculiarly liable to this abnormal condition. The hock joint is in a high degree susceptible to this change, and I have often made dissections and found all the bones of the hock, except two, strongly united together by this bony union, and in which not the least particle of motion could take place except between the astragalus and tibia. The inflammation which precedes this condition is very frequently the hidden cause of lameness, and that of a very chronic character. This condition may be accompanied by an outside swelling (Bone Spavin), or it may not. If there is no outside swelling, there is no way of making a positive diagnosis of this condition. Therefore any lameness in the hind limbs that is of a chronic character, and cannot be definitely located, should be a sufficient cause to prevent the animal from being sent to the stud for breeding purposes.

**BROKEN WIND.**—This disease is well known to be inherited, and any animal suffering from it should be avoided by the careful breeder.

**CATARACT.**—This is a chronic disease of the eye, and can be easily discovered by looking into the pupil. If cataract is present there will be noticed a light colored speck of varying size, from that of a pin head to a spot large enough to fill up the entire pupil. This disease is inherited beyond a doubt, and is a frequent cause of blindness, and if a breeder wishes to obtain animals with perfect sight there should be no horse with a cataract allowed in the breeding establishment.

**CURB.**—This is a disease that very frequently causes lameness. When present, even if the animal is not lame, the selling value is very materially lessened, even in a common grade of horses. How much more would be the depreciation if it was a cause of lameness, and thereby decreased the speed of a racing animal, the animal being probably made worthless. I have known a sire affected with curb to get colts, three-fourths of which were affected with this disease before they were six years old. In fact there are numerous instances showing the strong tendency of this disease to be inherited. Never breed from a sire or dam that has a curb.

**DIARRHEA.**—There is a form of diarrhea which is sometimes present in trotters, that seems to be caused by a constitutional irritation of the bowels. In many cases the subjects of this affection cannot be driven in a speedy manner for any length of time without causing diarrhea. Such horses are very liable to become debilitated, and easily the subjects of chills and colic and to lose their appetite for twelve or fifteen hours. Such animals are often said to be delicate and unable to stand any severe exertion, when frequently repeated. An animal that is subject to this disease should not be bred from.

**HEMORRHAGE.**—Many thoroughbred horses, when made to run to the top of their speed, are affected with this serious condition of bleeding from the nostrils. This is also due to inherited and constitutional causes, and if this condition is present in any proposed candidate for the stud, he or she should be rigidly excluded from the breeding establishment.

**FEET—IMPERFECTIONS OF.**—The shape and form of the feet tend very strongly to be inherited, and any weakness which may be present in the sire or dam will be very likely to appear in the offspring. One of the worst faults in respect to the feet is seen when they are small and cramped, and when the horn is thin and has a tendency to become dry and brittle. This form of hoof is especially liable to corns and quarter cracks and are the easy subjects of bruises to the soles, by stepping on stones and other solid objects projecting

2

from the surface of the road. I have known a stallion which had such feet to reproduce colts about half of which could not be driven on macadamized roads for any length of time without going lame from bruises to the soles, or a cracking of the thin, dry horn at the quarter of the foot. A large foot is not so objectionable from a utilitarian point of view, but may be somewhat objectionable in race horses. It may be said, however, that a large foot is seldom to be considered an objection in an animal that is otherwise suitable for breeding purposes. Many horses that travel apparently sound are seen with contracted heels, and although this is frequently due to bad management of the feet, there are many cases in which horses are predisposed to this condition from thin, weak horn. When it is proposed to breed from an animal with contracted heels, and the cause cannot be plainly traced to bad management of the feet, including bad shoeing, the subject should be excluded for breeding purposes. Be sure that both dam and sire have well formed and good sized feet.

**NAVICULAR DISEASE.**—Although this is not such a common disease in this country as it is in Great Britain, in consequence of our having softer roads, yet it is frequent enough, especially in animals past middle age that have been driven for any length of time in the vicinity of towns, where there are stone pavements or macadamized roads. Do not breed from any animal that is the subject of navicular disease. All authorities agree that it tends to be strongly inherited.

**OPHTHALMIA.**—In March, 1893, a gentleman sent, for my examination (requesting a written report), a beautifully formed cob mare that was then suffering from an attack of inflammation of the eyes, and had been the subject of another attack two months previous. The eyes were clouded over, showing all the appearance of a constitutional ophthalmia (moon blindness). He wished me to report especially on the advisability of placing her on his stock-farm for breeding purposes. Knowing as I do how strong is the tendency for eye diseases to be inherited, I was compelled to advise against the step he contemplated. I should have been exceedingly derelict in my duty had I done otherwise. Whenever an animal is the subject of cataract, or cloudiness on the front of the eye—thus showing the effect of ophthalmia—it should be rigidly excluded from the breeding farm.

**OSTITIS or SORE SHINS.**—This may be considered almost exclusively a disease peculiar to thoroughbreds, and from the fact that many otherwise good horses break down when young from this disease, they are very likely to be sent to the stud for breeding purposes. Now, inasmuch as the very fact of their having broken down is a proof of constitutional weakness in the bones of the limbs, it should be a sufficient cause to prevent breeders from getting more of the same kind of weak-boned weeds that are more than likely to turn out to be useless as racers or for any other purpose. Beware of an animal that shows the sign of having been at any time affected with sore shins.

**RHEUMATISM.**—Although rheumatism as a disease may not be directly inherited, the constitutional condition which makes an animal peculiarly subject to it is probably inherited. I should therefore strongly advise against breeding from an animal that is the subject of rheumatism.

**RICKETS.**—This is a constitutional disease and affects animals while they are yet very young, The peculiar condition of the bones which causes them to bend or give way under the weight of the body has a strong tendency to be inherited. It should therefore be the aim of the breeder to pick out animals that have no appearance of having the long bones bent in early life.

RINGBONES.—All authorities recognize that the tendency to ringbones is strongly inherited. It is therefore very necessary for the successful breeder to keep all animals affected with ringbones from becoming members of his breeding family.

ROARING. —Youatt gives some strongly convincing instances where sires affected with this disease had gotten numerous progeny that were the subjects of roaring. He says: "Facts have established the hereditary predisposition to roaring beyond the possibility of doubt." A well known owner of horses said to me that he dreaded an outbreak of distemper among his horses, as it was almost certain to leave some of them as roarers. There is no doubt in my mind that if breeders were more careful in this respect thoroughbreds would not be more subject to be left roarers after distemper than are other breeds of horses. The rule should be to never breed from a roarer.

SIDEBONES.—This is often a disease of the more common bred horses, but nevertheless is strongly inherited and is a cause of making numerous animals comparitively useless except for very slow work. The breeding of animals affected with sidebones can only result in a deterioration of the quality of horses when considered in the aggregate.

SPAVIN.—This is considered by all authorities to be a disease which is very strongly inherited, and as it is a very serious affection and a great source of loss to horse owners, it would seem to be the height of folly to breed from animals already diseased, and which are almost certain to reproduce in their progeny the tendency to take on the same abnormal character.

While visiting in the country some years ago, a patron of mine called my attention to a foal about three weeks old that had an enlargement on the inside of the hock which I quickly found to be a true bone—spavin. I inquired carefully about the sire and dam. I could not find out anything about the sire that could be of value in coming to a correct conclusion regarding the condition of his hocks, but the owner of the colt still owned the dam. On examination I found that she was a plucky cob that had been worked hard to a butcher's cart for years; that she had two well developed spavins on the hocks, which even then were a cause of the stiff and stilty action so characteristic of horses with fully developed spavins, and in which the inflammation—which is always present during the period of development of spavin—had previously disappeared. The owner informed me that he had noticed the peculiarity at the time the colt was foaled. I have never seen any but this case where a foal was born with a well developed spavin, but spavins frequently develop at a very young age, and in many cases it will be found that the progenitors were the subjects of the same disease. Therefore, in breeding horses of every class care should be taken that horses with bone-spavin should be excluded from the farm.

**STAGGERS.**—This is a very serious disease and no animal should be bred from that has at any time shown any symptom of it.

While I have used the above strong protest against breeding from diseased animals, it is not considered that the progeny of diseased animals will all bo affected; in fact a great many will probably escape, and this will depend to a great extent on the degree of prepotency that may be present in the diseased parent.

Lehndorff, a great authority on breeding racehorses, says : "The principal requisite in breeding a racehorse is soundness ; again soundness, and nothing but soundness."

TEMPER.—I cannot refrain from uttering a warning against the too frequent practice of sending bad tempered mares—that are uncontrollable in ordinary occupations—to the stock farm for breeding purposes with the remark that "she is not useful for work, but she will probably drop a good colt. There is nothing more certain than the fact that the temperament of the parent is very frequently reproduced in the offspring and every breeder knows what a useless article a colt with a vicious and uncontrollable temper will prove to be; in fact, he will be more profitable dead than alive, and will certainly be less dangerous to the lives of attendants that might be brought into contact with him. A good temper is one of the most valuable traits in the constitution of a breeding parent. This should be made a cast-iron rule with breeders of all animals.

# VARIATION.

WITHOUT a tendency to variation there could be no advance in organization or progress in results.

Variation is directly opposed to heredity, so that these two directly opposite forces are continually warring with each other for the mastery; and it is the object of the breeder to take advantage of any little change that may be to the benefit of the animal which he is attempting to improve. The improved animal does not always show it in his physical proportions. Especially is this true of thoroughbreds and trotters. It is impossible to predict in advance the career of any yearling colt, either runner or trotter, from his form or shape, and successful purchasers of horses at the breeders' sales depend principally on the breeding of the animals they buy, trusting that the force of heredity will be stronger than the opposite force of variation. Yet we very frequently find that the most carefully bred colt has varied sufficiently in some particular character to make him comparatively useless for racing purposes. On the other hand, we occasionally find that a colt that has been what may be considered rather carelessly bred, will show enough variation from his parents in the right direction to be an extremely valuable animal. The tendency to variation is the force which is the cause of those two anomalous facts. There is one quality in the composition of horses and dogs that is probably of more importance than any other, and is entirely hidden in a superficial view of an animal. I refer to the nervous organization. It is to this part of the animal system that is due the much greater strength and endurance of some animals when compared with other animals which resemble them in external appearance. Every horse owner knows that we may take two horses that are almost exactly alike in size, shape and action, that one will turn out to be a very valuable animal for work, while the other, though subjected to exactly the same conditions as to management, will be of very little practical utility. The difference is due to the better nervous organization of the useful horse. We may also say that the nervous system is also subject to variation in a high degree, and that this is a factor that cannot be discovered by any observation of the form of the animal. This fact will always make the breeding of fast horses more or less problematical and in consequence of its uncertainty, give it a speculative character. It is by taking advantage of variations in form or action that the different varieties of each species of the domestic animals have been selected and have now become distinct breeds.

In this short work, it will not be possible to go into details of the best methods for breeding thoroughbreds, neither is it necessary, as there are many books written by men who have had far better facilities for observation than myself, and who have written very fully on the subject of pedigrees and breeds of successful race-horses. There are a few general hints that may be inserted for the guidance of those who have not the facilities to obtain the more expensive treatises on breeding.

CLIMATE.—This has much to do with the results of breeding. It may be said that steep hills, marshes and low lands are not suitable for the successful breeding of horses. In countries where there are periods of prolonged cold, the conditions are opposed to the breeding of fine horses. Intense cold tends to stunt the growth of the young animal. Young animals should be well fed if we expect to obtain well-developed, full-grown adults. There should be no over-crowding on the stock-farm.

It is not sufficient that a sire should possess the formation it is desired to correct in the dam, or that the dam should have qualities likely to improve defects in the sire, in order to insure the obtaining of a perfectly formed product, but both sire and dam must be well shaped to get progeny that may be better than the parents. Young or middle-aged dams usually bear animals that have more vigor and stamina than old or extremely young dams.

William Day gives the following advice for the selection of thoroughbred mates : "Consider carefully the external form of the mare, the relation of different parts to each other, her capabilities, so far as known ; above all, her breeding and that of her ancestors; then select a sire on the same careful system." It will only pay to breed from the best stock, no matter what kind of domestic animal it may be. This applies equally to horses, cattle, sheep, pigs, chickens, etc.

It is usually considered that soils resting on a chalky or limestone formation are the best suited for the location of breeding farms.

## ABORTION.

This is a very important subject to breeders of high-grade animals and in the aggregate, a source of immense loss. Billings says : "that it causes an immense loss to the agriculturist and breeder." In some statistics collected by the New York commissioners the average number of abortions was about 5 per cent. of the whole number of cows that were pregnant. Many years ago it was estimated that the loss in New York State alone was over four million dollars a year. In mares it is not of such frequent occurrence as in cows, but still frequent enough to cause considerable loss.

*Causes.*—The facts point to the certainty of an infection frequently being the cause in cows, as the following will show : A German authority reports that all the cows aborted in one stable, while none aborted in another on the same farm. In another case abortion continued after every possible change in the manner of feeding. Other cases are reported where it has existed for a number of years, constantly increasing until finally nearly every pregnant cow aborted. In another stable one cow after another aborted, while none occured in a second stable on the same farm, until a maid who had assisted at an abortion at the first stable returned to the second stable and attended the cows there, when abortion set in and continued for a long time. These and other facts point to the absolute certainty of the infectious nature of the malady.

It is very frequently caused by violence in the mare, and in fact this is a very common cause in all animals. Falls, kicks, excessive labor, great exertion, any of the violent inflammations of the internal organs, irritant medicines, and diarrhea are all causes of this affection. I have known it to follow the casting of a mare for an operation, also from decaying animal matter, as the refuse from a slaughter house. A friend of mine, a butcher, informs me that he has often tried to breed mares that are stabled near his slaughter house and always without success.

The following medicinal agents are said to be a cause of abortion : Cantharides, Tansy, Savin, Cotton-root bark, the various forms of ergot, and probably other fungoid bodies that are frequently found in musty fodder.

# DURATION OF PREGNANCY IN THE DOMESTIC ANIMALS.

| | DAYS | AVERAGE DAYS | |
|---|---|---|---|
| **MARE** | 320 to 365 | 340 | 11 months and 1 week. |
| **COW** | 240 " 300 | 283 | 9 months and a few days. |
| **SHEEP** | 143 " 156 | 149 | 5 days less than 5 months. |
| **GOAT** | | 160 | 5 months and a few days. |
| **PIG** | 110 " 130 | 119 | 17 weeks. |
| **BITCH** | 58 " 65 | 63 | 9 weeks. |
| **CAT** | 50 " 64 | 55 | 8 weeks. |

# PARTURITION—FOALING—CALVING, ETC.

THIS is the act of the normal expulsion of the mature fetus. In the higher animals it is a very complicated physiological process, and fraught with considerable danger to both the dam and progeny. And the more we deviate in our standard of breeding from the original wild type of the animal, the greater will be the danger of fatal accidents attending parturition.

Fig. 4.  Cow in the act of parturition.

*Symptoms.*—The preliminary symptoms or signs of approaching parturition are an increased size and sensitiveness of the mammæ (udder.) The tenderness increases until the fetus begins to feed on the milk of the mother. The vulva becomes swollen and flabby. This is followed by restlessness, the mother lying down and getting up again in much the same manner as in colic, and she often seeks a remote place to bring forth her young. The pains now become more severe ; they are more frequent, and they continue for a longer time; this increases until the fetus is expelled, if it is a normal labor. The naval cord is now ruptured and there escapes perhaps a little blood. The time of expulsion of the fetus after the actual pain varies in different animals, but if the labor is normal the mare will expel it in from five minutes to half an hour after the first actually visible symptom of pain. The cow from half an hour to two days. Sheep from fifteen minutes to three hours.

Fig. 5.   Mare in the act of parturition.

**EXPULSION OF FETAL MEMBRANES OR AFTER-BIRTH.** This may occur soon after birth, or it may be delayed for a variable period. It is unusual for the after-birth to be retained for more than an hour or two in the mare, but in cows it is frequently retained for several days. It may be generally said that there is no great need of removing the retained membranes by mechanical means for a week or more after birth, unless there is a high external temperature of the atmosphere, or unless the genital organs are abraded, or if there is frequent straining, and especially if there are fetid or bad smelling discharges from the vagina. In these cases it is necessary to remove the decomposing membranes as soon as possible.

Fleming advises that when necessary the retained membranes be removed from the uterus of the cow by passing in the hand and gently tearing away the membranes from their attachments about the third day. However, it is often possible to remove them by gently pulling on the small string of membranes that usually projects from the vulva and hangs down.

The following formula for a medicine to assist in the expulsion of the retained membranes is recommended by Hering.

| | |
|---|---|
| Carbonate of Potash - - - | half an ounce |
| Savin leaves - - - - - | one ounce |

These to be infused into one pint of water, then filtered, and given warm in conjunction with linseed tea, and repeated every six hours until the membranes are expelled.

## INVERSION OF THE UTERUS.

It sometimes happens that the uterus is turned inside out as it were, and the whole organ will be protruded from the vagina, and hang in a pendulous manner when the animal is in the standing posture. When in this condition an attempt should be made to return the organ. The following method will usually be the most successful.

If the animal is in the lying position, she must be made to rise if possible, as it will be much easier to return the organ when in the upright position. Also attempt to rig a contrivance whereby the hind quarters will be raised higher than the front part of the body. Now take a bed sheet or something similar and fold it double. Place assistants holding two corners of it each side of the uterus, and place it under the organ. Now have the men raise the uterus. The operator, standing behind, should begin to return the organ to the abdominal cavity by pushing carefully the parts nearest the vulva, and gradually working more and more back into the cavity until the whole of the uterus has been returned. The animal will strain considerably at times while this is going on, and it will be necessary for the operator to use continuous pressure to prevent the ute-

Fig. 6. Loop of rope to form Delwart's truss.

Fig. 7. Delwart's truss applied.

rus from returning again to the outside. But it can be accomplished successfully if the operator uses patience and does not for a moment relax his vigilance. After the return there is sometimes considerable straining, and it may be necessary to keep in close proximity for a few minutes to prevent the uterus from being forced out again. The organ being in place again it is usually necessary to take steps to prevent another inversion. I have found the following to be very successful: take a large bag needle and thread it with strong tape. Insert two stitches, allowing them to cross each other. They should be inserted rather deeply so as to secure a good hold. In this way a return of the organ or even a portion of it very rarely takes place. Another method to prevent a re-inversion is by the application of Delwart's truss, which can be made with an ordinary rope, (figure 6) and applied as in figure 7. Several other contrivances have been used, all having the same end in view, namely narrowing the outlet from the vagina. None of these appliances should be placed so tightly that the animal cannot pass the urine.

Many seemingly hopeless cases of inversion may be remedied by using the above means. I will give one instance to show the recuperative powers of the organ. Being sent for by a breeder, six miles from my residence, I arrived at the farm in a storm of sleet and rain which had lasted several hours. I found the cow in the field, lying down with the whole uterus protruding. It was covered with darkish purple spots and cold, and to all appearance had no chance of being restored to a normal condition again. I had a sheet brought, and with the necessary assistants we raised the cow and lifted the uterus with the sheet; then led the cow to a shed, where I operated as above explained to return the organ, and placed two stitches across the lips of the vulva. The uterus was retained and a complete recovery took place in a few days. The uterus retained its normal position.

3

## PRESENTATION OF THE FETUS.

By presentation is meant the part of the fetus which first presents itself to the uterine outlet.

In Fig. 8, is a representation of a twin pregnancy from Fleming.

Fig. 8.  Position in normal twin pregnancy.

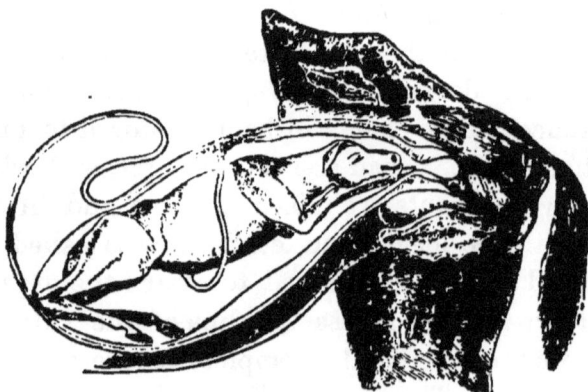

Fig. 9.  Normal presentation.

The most common way for the fetus to be presented is represented is Fig. 9, and is usually considered its natural position. As will be seen, the nose and fore feet are presented first. With this method of presentation there is rarely need for artificial assistance.

Fig. 10.

A presentation of the hind feet is seen in Fig. 10. In this case delivery is usually effected by natural forces of the mother without outside help.

There are many cases of what may be termed false presentations where the greatest skill and ingenuity of the operator will fail to make a successful delivery. There are other causes which may prevent a normal delivery, but they can only be mentioned in this place and the more common cases slightly touched upon.

The cases of difficult labor are from ten to fifteen times more frequent in cows than in mares. They are also very common in bitches, caused to a great extent by the fact that large males are frequently mated with small females, in which case the fetal animals will probably be larger than the mother can expel under normal conditions of labor. Difficult parturition is much more dangerous in mares than in cows. It is said that if labor is delayed more than three or four hours in the mare the death of the fetus is sure to result, whereas in the cow it frequently happens that parturition is delayed from 24 to 48 hours without any disastrous result to the fetus. However, where labor is delayed over a reasonable time, no delay should be allowed, but an attempt made to discover in what position the fetus is placed. In order to do this the hand should be well oiled, and inserted into the vagina, and if the opening into the uterus is sufficiently dilated, the hand should be carried into that organ also. By this time some of the fetus will be felt. If it is a limb, an attempt should be made to discover whether it is a fore or hind limb. This will be easily determined if the hand is carried as far up the limb as the knee or hock. If it is a hind leg, the joint of the hock will be felt. It should also be remembered that the hock joint bends in the opposite direction to that of the knee. If it is a fore leg, an attempt should be made to find its mate, and also to try and feel the head, and discover if the nose is pointing toward the vagina If it is a hind limb, an attempt should be made to find the other and feel if the tail is coming toward the uterine opening. While making these explorations it should also be noticed whether it is an extraordinarily large fetus or not, or anything that may be peculiar about it. And here I may mention that in my practice I have found one case in which the uterus was twisted as

seen in Fig. 11. In this case it was impossible to introduce the hand or even two fingers. Of course the case
terminated fatally. This impediment is rare.

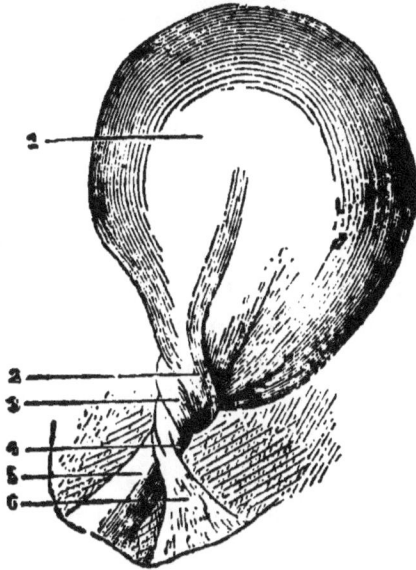

Fig. 11. Twisted Uterus.

It may be found that the fetus is very large as a
whole, or parts of it abnormally large, as in dropsy of
the head, chest or abdomen, or from emphysema ; that
is, where decomposing or putrefactive gases have been
generated in the tissues under the skin of a dead fetus.
If there is found to be dropsy of any part of the body,
it may be necessary to puncture the cavity and allow the
fluid to escape, when the fetus will more nearly assume
its normal size and probably allow of a fairly easy delivery. If there is general emphysema, it will be detected by the skin having an elastic feel when pressed.
When this is the case it will often be necessary to make

numerous punctures through the skin. This can be done by a knife or any pointed instrument. Where this is present, it will be found that the walls of the uterus and vagina will be very dry and it is necessary to use oil or grease freely before an attempt is made to deliver a fetus of this description.

In this place it may be also mentioned that monstrosities are sometimes found. These vary in numerous ways, which are exceedingly interesting to the embryologist, but which cannot be noticed here. I can only give a short description of one of the most interesting as well as a very rare specimen that I met with several years ago when practicing my profession in an agricultural neighborhood. On examination I found three feet coming almost together. Passing my hand along the legs, I found that I had two hind feet and a fore foot. Groping further I soon felt what apparently was some of the intestines and internal viscera of the fetus. Being all at sea as to what it meant, I followed up the fore leg and tried to find the head, in which I was unsuccessful, neither could I find the other fore foot. I expected to have a difficult labor, so attached two ropes to the two hind legs and gave them to assistants. After using a quantity of oil to lubricate the parts, I ordered the assistants to pull gently but firmly, while I attempted to guide the fetus so that the hind quarters would come through the outlet all right. By strong and steady pulling, together with the efforts of the cow, a successful delivery was accomplished in about ten minutes, when the following fetal condition was seen. The abdominal and thoracic cavities were open, the walls of which were turned back, much like the abdominal walls of a lamb when killed and dressed for the market. The diaphragm was in place and the viscera in their normal situations, but the neck was doubled backward and the head lay back on the dorsal portion of the spine, which explained why I had been unable to find it on making the manual examination. I believe the fetus had been alive up to the time of my manipulations, but probably died while coming through the outlet from pressure or perhaps some other cause. At any rate it did not breathe after being born. The cow progressed in the normal way after delivery. We will now consider a few of the more common presentations in which assistance is required,

If there is a normal presentation and the water-bag has ruptured, the operator can frequently cause a complete delivery by simply oiling the parts, taking hold of the legs above the feet, and pulling strongly at the same time that the animal has the pains. In this manner the fetus comes nearer to complete expulsion at each period of labor pains. If it cannot be accomplished in this way, ropes or strong cords should be applied above the feet and given to assistants. The operator, passing the hand up, tries to place the head in an easy position for exit, and he may at the same time use traction on one of the jaws, while the assistants exert traction on the cords or ropes at the time of pains, thus assisting delivery.

**FETUS WITH FORE-LIMBS BENT AT THE KNEE.**—Figure 12. Sometimes one knee only is bent, at other times both. This is usually a difficult presentation to remedy unless the attempt be made early. The hand should be passed back alongside of the neck and the limb grasped just above the knee and pushed back, the shoulder elevated toward the ears of the fetus ; then secure the leg at the feltock and try to bring it toward the outlet. If both knees are bent, this should be repeated to get both feet near the opening. Having done this, deliver in the way described for normal presentation when difficult.

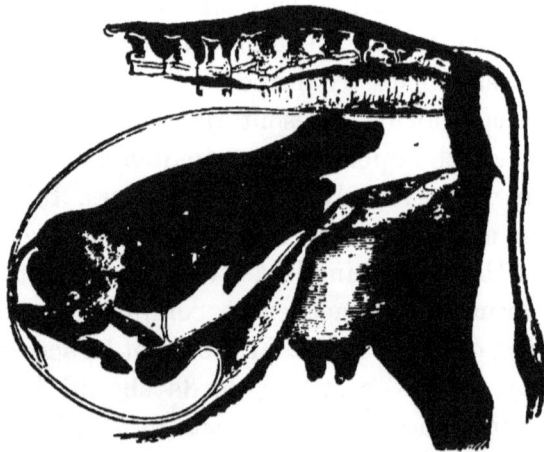

Fig. 12. Fetus with fore-limbs bent at the knee.

**FORELIMBS COMPLETELY BENT BACK UNDER.**—Figure 13. In this presentation push back the foal or calf as far as possible into the uterus, and bring the limbs gradually toward the outlet, as recommended in the last presentation (bent knee). Then deliver in the same manner as normal presentation.

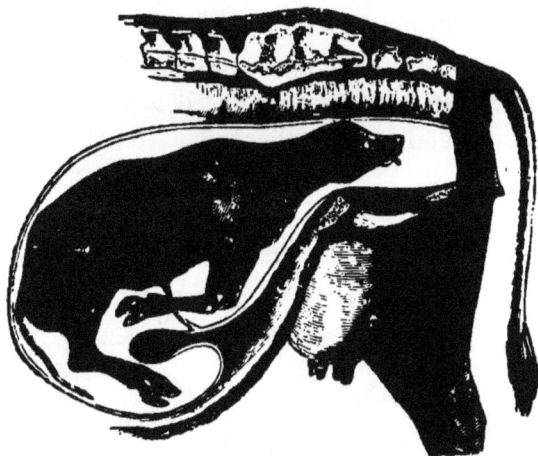

Fig. 13. Fore-limbs completely bent back under.

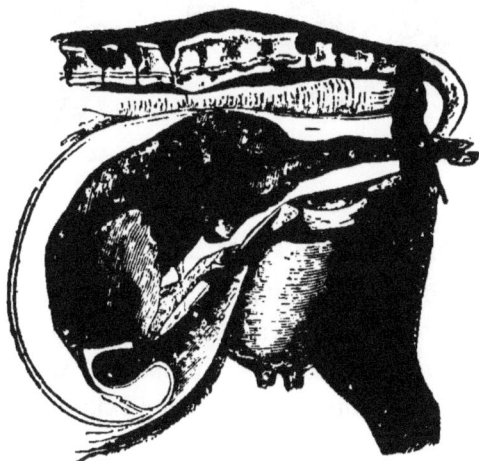

Fig 14. Head of fetus turned downward.

## HEAD OF THE FETUS TURNED DOWNWARD.—

Pass the hand along toward the nose of the fetus; grasp it and lift the head so as to be able to turn it toward the opening, and deliver as before recommended.

If there is an upper deviation of the head, an attempt
must also be made to bring the nose toward the outlet
by bringing the head around. If it is sideways the same
method of procedure is indicated.

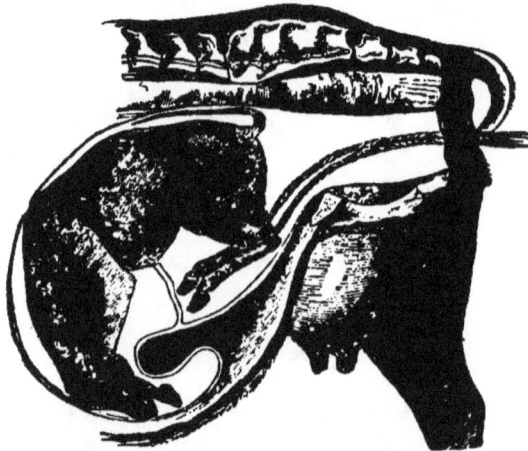

Fig. 15. Hock presentation with cord attached.

**HOCK PRESENTATION.**—Push the fetus as
far as possible into the uterus. This will be materially assisted by raising the hind quarters of the dam
above the level of the fore limbs. If possible, turn
the feet backward and let the labor take place by hind
foot presentation. If necessary, to complete delivery
ropes can be attached to the legs and traction applied,
as already explained, by assistants under the direction of the operator. If the legs cannot be turned
back in this manner, ropes may be applied, as shown
in the figure, the assistants bringing the limbs as far
out in the vagina as possible, when the operator can with

comparative ease cut off the limbs at the hocks, or at any rate divide the tendons behind and above the hock, when the limbs will easily bend up double, with the feet close to the styfle. The delivery will usually be effected with comparative ease.

I have not mentioned the various instruments used by veterinarians to facilitate delivery, as in most cases it would be dangerous for an ordinary operator to attempt their use. In all cases where the limbs are turned back and cannot be easily brought around, there should be no hesitation in using a great deal of force to push back the body of the fetus, to allow the limbs to be turned the more easily. The principles to be observed in effecting delivery are purely mechanical, and allow a considerable degree of latitude in the selection of appliances to aid the natural powers of the mother.

## BREAKING AND TRAINING OF COLTS.

IT is a very simple matter to break and train colts when the handling is begun at a sufficiently early age and the earlier the education of the future horse begins, the easier will be the task. However, on account of the breaking taking up more time than the ordinary breeder is willing to give to it at an early age, it often occurs that colts are not broken until they are two or three years old. The breeder has now a very difficult problem to deal with, for if he makes a serious mistake, it is likely that his colt will turn out a run-away, a kicker, or evince his vicious propensity in other ways. We will therefore consider first, the proper steps to take to properly bring up a colt so that he will be a quiet, tractable animal, and therefore very desirable to own.

It will be found that on account of great difference in disposition, temperament, tractability, docility and intelligence, colts will require various means to educate them to become useful horses.  Some are naturally stupid and difficult to teach ; others are very nervous, and everything strange alarms them; others are stubborn and perverse ; each requiring to be handled according to its temperament and tendency.  The two governing principles with the handler should be kindness and firmness.

Begin to break the colt to the halter when a few weeks old, and get him so that he will lead easily. Run the hands down over his legs occasionally and after a time lift his fore feet one after another for a few moments.  In this way you will get him used to being handled in various parts of the body without any fear of his being harmed.  Having done this, it will be a great help to the horse-shoer when he is taken to the shop to be shod for the first time ; and he should be taken at a time when no other horses are around, or only one or two, so that he will not be apt to get restless and excited while waiting to be shod, for, on account of the strange noises made by the hammer on the anvil, etc., colts are often wrought up to a high pitch of excitement previous to the beginning of the shoeing operation.  It may be well to give spirited colts a considerable amount of exercise before taking them to the shop, to make them more tractable to the manipulations of the horse-shoer.  Neither should any colt be placed in the stall alongside of a kicker, cribber, windsucker, weaver, or horse with other visible vices, as they are often imitated in an incredibly short space of time.  I have known a cribber placed alongside of another horse for only one night and the other horse to become an incurable cribber within a few days afterward. Young colts should not be overworked, as it often causes them to become apathetic, lazy, and sometimes vicious, besides being a cause of various diseases.

Fig. 16. Leg strap.

Fig. 17. Horse showing leg-strap applied.

We will now consider what is best to be done with a colt that proves to be unmanageable, when ordinary means are used to get him broken. We will suppose that the breaker has succeeded in getting on an ordinary halter, but that the colt refuses to be led, or has broken away from the man as soon as he gained the open air. He must be returned to the stable again. At the end of the halter should be tied an extra rope about 20 or 30 feet long. The breaker should take in his hand a rope about 5 feet long with a loop in one end, or what is better, a stirrup strap or breeching strap of a single harness, or the leg strap seen in Fig. 16, and advancing carefully alongside the colt or vicious horse, pat him gently on the neck and shoulders for some minutes, gradually but steadily working his way with his caresses toward the fetlock of the fore leg. Now attempt to lift the foot and put it down again, repeating this operation 3 or 4 times, each time keeping the foot suspended longer than before. Having to some extent gained his confidence, try to pass the rope or strap around the fetlock, and when the foot is lifted again, secure the strap by tightly buckling around the arm. If a rope is used, it should be secured by a half-hitch bow-knot, so that if the necessity occurs the rope can be slackened immediately, which could not be done if a proper half-hitch knot were used. Now the horse will begin to plunge and move around lively— even if he is on three legs—and the breaker should keep at a respectful distance, with a good hold of the rope at the end of the halter shank, trying to steer or drive him toward the door into the open yard or lot. Of course it will be best to get him on to a field where the ground is soft, although there is no danger of the horse injuring himself when restrained in this manner, but sooner or later he will probably lie down, and as it will be best to give him a large amount of 3-legged exercise, let him run as much as he is inclined to, occasionally giving him a sudden pull around by tugging strongly at the halter shank, just to let him feel that he is under control. After giving him time to become a little fagged, gather the rope up slowly toward the head, and as soon as he will allow it, pat him on the neck and body, gradually working your way so as to handle the hind quarters and limbs. If he suddenly starts up and tries to jump away from you, give him some more running around, until he will let you handle him all over. He should be thoroughly tired out before being allowed to get on all four feet again. He must also be willing to allow the breaker to get around him without showing fear. Now let down the foot, and try to lead him around gently for some time, which will usually be accomplished without trouble. I have never yet found a horse that I have been unable to secure in this manner, although I have handled a large number of wild mustangs and other vicious horses. It may be necessary to repeat this operation the next day and for several days afterward, if the animal is stubbornly vicious. I believe it is possible to tame any horse by this means if it is persevered in for several days. Another method for securing a vicious animal, and in fact any horse that is liable to be dangerous to the attendant, and also one that will be very useful when having to perform small operations on the quarters or hind limbs, as sewing up wounds or firing, is to place a strap on the front leg, as described above and shown in Fig. 16, In addition apply a side-line, to the hind fetlock, as shown in Fig. 18, having the rope brought around to the front and held by an assistant. This will also prevent the horse from rearing. It will be found that no matter

how vicious the horse may be he is comparatively pow-
erless for mischief when this is applied, and the operator will be quite safe even when placed behind the
hind limb.  These will be all the special appliances
necessary to be used in breaking even the most refrac-
tory animals.

Fig. 18.   Hind-limb secured by a sideline.

It is well in this place to introduce a common, and
it may be said, a very convenient method for throwing
horses, a method that can also be used on vicious
horses when the necessity for throwing them arises.  A
drawing of this is seen in Fig. 19.  As will be seen, the
only special apparatus required is a strong rope about
25 to 30 feet long, such as can be found on every farm.
This should be doubled, and about two or three feet
from the bend a knot is tied, as seen at A.  This forms
a loop, which should be large enough to encircle the
neck like a collar.  It will be well now to strap up the

fore foot as seen in Fig. 16 previously described
Then, having chosen a soft place, carry the ropes back
under the breast, between the fore legs and between
the hind legs around below the fetlock, then bring
them forward, one on each side, and through the loop.
Two assistants should take the rope on each side with
one man at the head.  By pulling on the ropes the animal will soon be taken off his legs and fall down, when
the man at the head should immediately place the
weight of his body on it to prevent struggling.  The
ropes can be securely fastened and the casting completed at the will of the operator.

Fig. 19.  Rope arranged for throwing Horses.

This will be very convenient as a means of restraining
animals on farms at a distance from towns where hobbles
might be obtained, as some horses will not allow even
the stitching up of small wounds without being thrown.

4

Second. Having got the horse or colt under control for leading purposes, it will be well to put on a bridle and get him used to having a bit in the mouth, wearing it for some time each day, and gradually get him used to being led by the bit. Care should be taken that the bit is not buckled too high in the mouth ; also avoid sudden jerking, as it is apt to bruise the gums, which are now very tender. This is probably the most frequent cause of bad mouths. If farmers could only appreciate the enormous losses in value that occurs in the best grade of horses through this circumstance, I am sure they would be more careful than they now are to keep the mouth in as perfect a condition as possible. If they will once grasp the fact that their best interest centres in breeding horses for carriage purposes, instead of raising them with a little speed and less size, in the hope that lightning will strike them and make fast trotters of them, they will be in a fair way to get a good price for them. In order to get a good price for a horse for carriage driving he must be of good size, well formed with a good head and neck, a good mouth and sound. A breeder need not look for customers for such stock, for dealers in fine horses are continually scouring the country in search of them. It is really surprising to see the large number of otherwise valuable carriage horses that arrive in New York every year, with spoiled mouths, which by bad training become pullers and have to be sold at an uncommonly low figure to be used at some common every day service. Now, if it is proposed to make a harness horse of the colt, the harness should be placed on him and allowed to stay on for some time, while the attendant handles it in various ways—pulling at the traces and tugs, tightening the breeching, etc. This should be repeated a few times before hitching up the animal. The colt should now be placed in a two-wheeled cart, and a kicking strap applied. It is usually good policy to allow the colt to stand still until he moves off of his own free will ; then to move along slowly being gradually made to feel the bit and to go slower or faster by gently tightening or slackening the reins. This process to be repeated every day until he is quite tractable to all changes of vehicle or grade of road.

In regard to the training of running or trotting horses, there can be no regular rule laid down, as each horse will have to get a special and individual treatment, in order to get the very best results from his training. In general terms it can be said that the most successful race horses are those that have a perfect digestion, and that have the power thereby of changing the stored-up force or energy that is contained in a large amount of food, into active and potential force, so that some of our most noted running horses take daily from 15 to 20 quarts of oats, and keep in the most perfect condition when taking even such large quantities. Then again some horses require from 8 to 12 miles of a canter while others will only stand a quarter of this amount, yet will win some very fast races for short distances. The rule is, however, that the animal that will stand the most severe work in training is the animal that wins large amounts of money. William Day, the celebrated English trainer of race horses, lays down the following valuable rule : " If a young colt that is trained or exercised along side of an old horse has not passed him when near the end of his trial, the old horse should be held up to allow the young colt to get courage and confidence in himself." This rule will also be advantageously applied to trotting horses. Mr. Day adds : "Many horses are made rogues by a violation of the above rule." and by the abusive use of whip and spur on two-year-old colts.

# MANAGEMENT OF THE HORSE.

**STABLE** —This should be roomy, with high ceilings and good ventilation. The temperature should be kept as near 60 degrees, Fah., as possible, but heat should not be preserved at the expense of ventilation. It is much better to keep your animals warm by extra clothing, than to keep them warm in a hot, close stable, with no adequate means of ventilation. A cold stable is also to be prefered to one that is too hot.

**FOOD.**—The principle of feeding, is to feed on such food and in such quantity and manner as will maintain the horse in the most perfect health possible, having regard to the service required of it. Some foods are much more easily digested than others. It has been calculated that only 20 per cent. of wheat straw is digested as compared with 76 per cent. of hay. A healthy, strong horse can digest much more food and quicker than a weakly one, and a sick horse may have its digestive powers seriously enfeebled, while hardship and fatigue have the same effect.

Food that has much dirt mixed with it may cause colic, indigestion, etc. Food that has become mouldy, from imperfect preservation or otherwise, is less nutritious and less digestible than clean food and is, besides, a frequent cause of colic, diarrhea, diabetes, skin diseases, paralysis and abortion in breeding animals. A sudden change from dry to green food often causes diarrhea, as also does new hay or oats, and especially is this the case with horses that are kept for fast work. On the contrary, a change of diet from green food to a dry and comparatively indigestible food will often cause constipation and indigestion, with other complications as a result. This is very frequently seen in the change of food that is given to horses when taken from the rural districts to large cities.

**COOKING FOOD.**—There are several reasons why the dry foods used for feeding the domestic animals should be cooked. First, is that of economy, the saving being from fifteen to twenty-five per cent. besides lessening the tendency to disease, and consequent loss thereby. It has also been proven that animals fed with steam cooked food, take on fat much faster than those that are fed on the same quantity of dry uncooked food showing that a considerable saving can be made in time required for fattening animals for market, and allowing the owner to dispose of them sooner than if dry fed, thus saving the feed that would be required to be given if kept feeding for a longer time.

Second. It would allow of the safe feeding of food that had been partly spoiled by being improperly cured—that is, mildewed—and containing the various fungi in large quantities. Musty hay and oats could be mixed with other materials and when properly steamed would be perfectly harmless and wholesome. Various apparatus are used, mostly consisting of an iron boiler with a connecting furnace and tubes for conducting the steam to the tubs containing the food to be cooked. It requires little or no attention and a very small amount of fuel will suffice to cook food for 100 horses. The price of the various apparatus ranges from fifty to two hundred dollars.

**GROOMING.**—The beauty, health and vigor of the horse are largely dependent on the condition of the skin. To harden the skin under a saddle or collar where there is apt to be very profuse perspiration, there should be applied, after taking off the harness, a diluted solution of HEARD'S DERMAL LINIMENT, as directed on the label.

## REMEDY FOR TEARING THE CLOTHING.

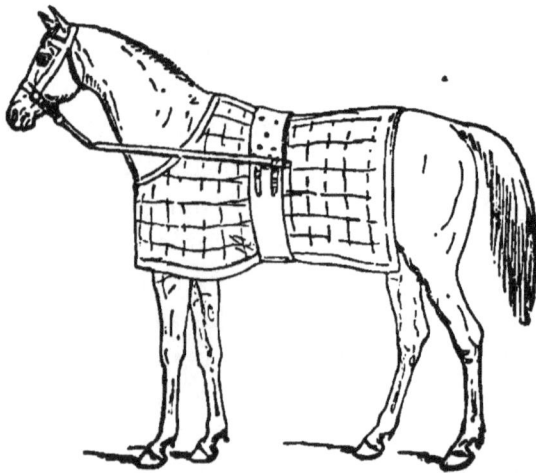

Fig. 20. Remedy for tearing the clothing.

IN figure 20 is shown a good remedy for the prevention of the habit of tearing the clothing.

**CLIPPING.**—This prevents undue perspiration in horses that are worked hard, and they are less subject to colds than horses with long coats. This is explained by the fact that horses that have long coats perspire very readily when working. The perspiration does not quickly evaporate through the hair, and the coat becomes saturated and hangs on the animal in much the same manner as would a wet blanket on a man. Experience has proved to me in a large number of instances that clipping horses is a great hygienic improvement on the old-fashioned way of allowing the old coat to remain on until spring.

**LOSS of KNEE ACTION.**—This sometimes occurs as a result of some acute disease. It is also frequently caused by working a young animal too hard, and I have frequently had my patrons say : "I gave one thousand or twelve hundred .dollars for that team, and when I bought it six months ago it had splendid action ; now it looks like a pair of old hack horses." On inquiry I have found that the horses which are probably only 5 years old, have been in the carriage for 3, 4, or 5 hours a day. The reason they have lost their stylish action is that the wear on the nervous system has been more than the supply of nerve producing elements would justify. This is also the case when good action disappears after sickness. And I may say here that this alone explains why some horses lose their speed after a severe attack of illness. It is my opinion that speed and fine action depend much more on the quality of development of the nervous system than it does on muscular development.

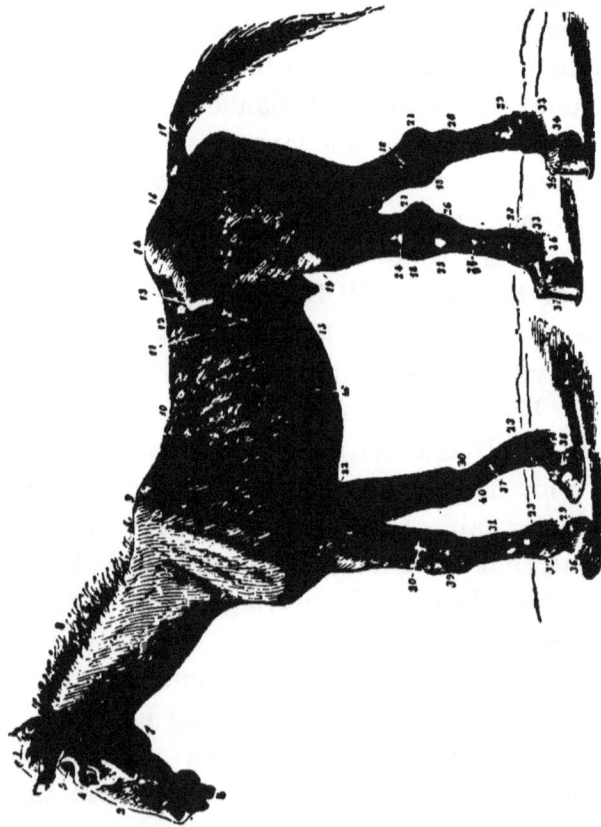

Fig. 21. Diseases and imperfections of the horse.

1. Glanderous discharge and ulcers.
2. Pendulous lips.
3. "Roman nose."
4. Cataracts, Amaurosis.
5. Hollow above the eyebrows.
6. Lop, or pig ears.
7. Glands or glanders.
8. Mange.
9. Fistulus withers.
10. Saddle-gall.
11. Tucked-up in front.
12. Weak-loined.
13. Ragged hips or haunches.
14. Mule, or goose-rump
15. Herring-bellied.
16. Flat-ribbed.
17. Rat-tailed.
18. Thorough pin.
19. Swollen stifle.
20. Knee windgall.
21. Capped hock.
22. Capped elbow.
23. Windgalls.
24. Callous tumour (Jarde).
25. Spavin.
26. Curb.
27. Simple splint.
28. Multiple splints.
29. Side-bone.
30. Weak tendon.
31. Sprained tendon.
32. Knuckling over at the fetlock.
33. Enlarged fetlock.
34. Grease or "Grapes."
35. Contracted tendons.
36. Foot deformed by laminitis.
37. Sandcrack at toe.
38. Sandcrack at quarter
39. Broken knee.
40. Bony tumours and swollen knee.

# PART II. DISEASES.

During a very busy practice as a veterinarian for the past 27 years, some of the early part of which was spent as veterinarian to the Third Avenue Street Railroad Co. (2,100 horses), the Bleecker Street Railroad Co. (600 horses), the Knickerbocker Ice Co. (400 horses), and the New York Transfer Co. (400 horses), I have naturally had the opportunity—seldom offered to veterinarians of the present day—to study in a practical manner the nature of all the various diseased conditions to which horses are subject, and to demonstrate by experiment the efficacy of the various means of treatment for the cure of each disease. This has been rendered all the more easy of accomplishment by my early training in the sciences of Chemistry and Animal Physiology, in each of which branches I received a government prize, given by the Science and Art department of the administration in Great Britain. Neither have I neglected the opportunity that this city affords of keeping up to date in the line of progress, having taken the special course in Pathology that is offered at the New York Polyclinic Post Graduate Medical School. Having done a large amount of work in Chemical and Pathological laboratories, I have had opportunities that fall to the lot of but few veterinarians.

In the few following pages I shall attempt to give the stock owner the benefit of my past experience as well as the best instruction that is offered by other authors in the treatment of diseases.

It is not my purpose to give a detailed account of the various diseases to which the domestic animals are subject, as that would require several volumes, each several times the size of this book. I shall aim, however, to give a short description of the causes, symptoms, the · best means of treatment, and the directions to be carried out, in language that may be easily understood by any one who is in the habit of handling stock. Neither shall I take up the reader's time by describing those rare diseases that are often only met with once or twice in the lifetime of very busy veterinarians.

To make it easy for the ordinary reader to find a description of the disease that his animal may be suffering from, I shall divide the subject into the following heads : Diseases of bones and joints, contagious diseases, diseases of the respiratory system, diseases of the digestive system, diseases of the urinary organs, constitutional diseases, injuries, abnormalities of teeth, and parasites.

## DISEASES OF BONES AND JOINTS.

### SORE SHINS.

This is a disease that the majority of running horses are subject to, and usually occurs soon after the beginning of severe training. In this disease we have an inflammation of the metacarpal or cannon bone and its fibrous covering.

*Causes.*—Violent concussion, such as the race-horse is subject to while the structures are in a young and tender condition and easily the subject of change, when conditions are unfavorable to the normal development of the bone. Winter training is particularly adapted to act as a cause of sore shins, on account of the severe concussions to which horses are subject when galloped on frozen ground.

*Symptoms* —Lameness in one or both limbs, after becoming very pronounced in a day or two, slight swelling, usually beginning on the front of the lower part of the metacarpal bone and sometimes extending upward, and to the sides. This swelling will have a doughy feel. There will be a good deal of tenderness on pressure being applied. When very severe there is some fever and occasionally a loss of appetite.

*Treatment.*—If the disease is discovered early, the leg should be placed in a bucket of hot water—as warm as can be comfortably borne by the hand —and renewed as often as required, to be continued for two or three hours, after which HEARD's AMERICAN EMBROCATION should be well rubbed in and a woolen bandage applied. This treatment to be repeated in twelve hours and continued until the skin gets somewhat roughened—the same as

Fig. 22. Metacarpal bone showing effects of sore shins.

in a very light blister. This will usually be much better than blistering, because the soothing influence is more continuously applied. When severe, it is also advisable to give a cathartic ball containing four or five drams of Aloes. The diet should be somewhat restricted.

In figure 22 is seen a drawing of the results on the metacarpal bone of a severe attack of sore shins. It will be seen that the bone is very much roughened from the extensive inflammation.

## SPLINTS.

Splints are tumors of the metacarpal bones. They are peculiar to the inside of the leg, and are usually situated from the middle of the metacarpal to two inches below the knee.

*Causes.*—In the evolution of the horse from three toes to one toe, the two side toes have gradually tended to become united to the center toe (see figure 23).

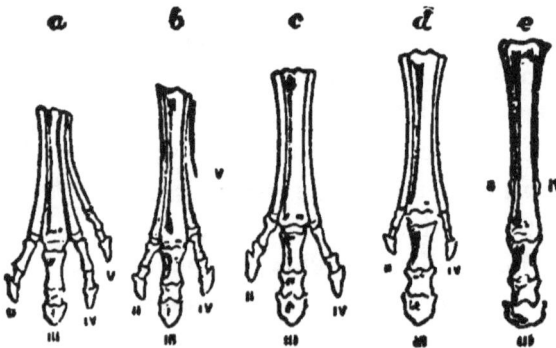

Fig. 23. Showing toes of the ancestors of the horse.

The transformation to the one-toed animal is not even yet complete, for the young horse has quite a respectable remnant of the two lateral toes still in existence. This is true also of fetal life. But it may be asserted that the horse is fast arriving at the stage where there will be no separate splint bones, and therefore no necessity for them to become consolidated in order to have a stronger toe to stand the great weight and severe strains and concussion of hard work.

Splints may be considered as additions of bony material to the leg, thrown out by nature to strengthen the limb that has been found too weak to withstand the labor required of it. The addition of this material is sometimes carried on so gradually that no lameness results, while at other times, especially when young horses are put to violent exercise, an intense inflammation is set up and great lameness is the result, which sometimes continues for several weeks, or until the process of deposition of the new bone is completed. We may safely say, therefore, that inheritance and active straining or concussion are the great causes of splint lameness.

*Symptoms.*—If they develop very slowly, it will frequently be noticed that splints will be found on the inside of the leg without the owner having observed any lameness or other symptoms that indicated disease of the limb. If it is caused by active concussion, there will often be considerable pain on pressure, and great lameness, which is more pronounced when going down hill.

*Treatment.*—When detected early, the horse should be taken from work and an application of HEARD'S AMERI-

CAN EMBROCATION made twice a day, rub in well and apply a piece of sponge or rag which has been soaked in the EMBROCATION, retaining it with a flannel bandage. In nine cases out of ten, this will effect a cure in from two to four weeks. In cases that resist these comparatively mild measures, recourse should be had to the firing iron, followed by a cantharides blister. I have found two or three cases in which on post-mortem examination, small bony tumors with sharp edges or points something like a knife, have grown out at the back of the metacarpal bone, and of course, right under the suspensory ligament, causing lameness whenever the animal was subjected to labor. In those cases it was impossible to discover the cause of the lameness, and it would have been useless to have attempted a cure even if the cause had been discovered. A horse that has perfectly sound action should not be considered unsound because splints are present; on the contrary, the limb is certainly stronger after the so-called splint has become fully developed, than t was previous to the beginning of the growth.


## SPAVIN.


In this place we shall treat of the diseases known as bone-spavin only, leaving the consideration of bog and blood-spavins for notice in another place.

It has usually been considered that the enlargement which occurs on the inside of the hock joint constitutes the disease known as bone-spavin. This narrow inter-

pretation has been a fruitful cause of dissension among
veterinarians, and a source of a great deal of litigation
between buyers and sellers of horses. It will be my
aim to make the description of the changes that take
place in the hock joint—as a result of disease leading
to the enlargement called spavin—so plain that the ordinary reader will have a tolerably clear perception of
them. As will be seen in Fig. 24, there are a number of small bones entering into the
formation of the hock joint (in the
young animal there are 9). Between
those various bones there is very little
motion at any time, yet it may be asserted that in the normal condition
some motion is allowed. The bones
are held together by a strong kind of
fibre—a kind of gristle. It quite
frequently happens that from some
cause or other an inflammation is
started in either the bones themselves or
in the fibrous structures which hold
them together. If this inflammation
becomes extensive, or is long continued,
one of two things happens : either
portions of the bone or bones ulcerate
and waste away, causing what is

Fig. 24. Hind leg
showing bones
of hock.

known as necrosis, or the opposite condition is the
result, viz: that there is new material of a bony nature
added to existing structures, filling up the space between the bones and of course destroying what little
motion is normally present. Now, we can have this con-

dition of new bone formation between other bones occur in any part of the hock joint. Sometimes its extension from the point of beginning to other bones in the joint is extremely slow. We can also have this bony material thrown out between any two of the bones, and as can be readily understood, without any extra enlargement outside of the joint. We may, in fact, have severe and long-continued lameness from inflammation or ulceration occuring all through the joint between the various bones, and yet have no external enlargement. This, then, is why there is such a variety of opinion on the subject of the presence or absence of spavin in horses that are lame in the hind limb. I have made a dissection in which all the bones entering into the formation of the hock joint were joined together as if in one bone, and in which there was only a small amount of enlargement on the surface of the bones where the so-called spavin is usually situated. In this place we shall call all cases spavin where there is an inflammation, or its results (which usually lead to a new bony formation) in any part of the hock joint, whether there is an enlargement on the surface of the bones or not. It can now be seen that the enlargement is apt to occur on almost any part of the surface of the bones of the hock. It is found, however, that in most cases the new growth is situated on the inside of the hock; very rarely on the outside. This is accounted for by the inside of the limb being more under the centre of gravity, and apparently having to take more of the concussion than the outside. Probably for the same reason the front of the inside is more frequently the subject of spavin than is the surface toward the back of the joint. Professor Williams says : "We can now understand why the external deposit is not the cause, but the result of the disease. So long as the ulcerated surfaces of the bones are unrepaired, the lameness will remain, but when the bones are united together (anchylosed) so as to form one bone and performing the functions of one bone, the lameness disappears, and the new material becomes as one of the essential structures of the body."

*Causes.*—Inheritance is perhaps the greatest cause. Hard work at an early age is also a frequent cause. In young horses, where the growth is not yet completed, the natural condition of the structures of the body is very easily upset, and diseased characters arise. It is also found that horses with small hocks are more frequently the subject of spavin lameness than those that are known as coarse hocked. Sprain of the fibrous ligaments, situated between the bones of the hock, and concussion of the bones—setting up an inflammation in the bones themselves—are the most frequent active causes. External violence may also be a cause. In many cases animals are foaled with one hock formed somewhat different from the other, and it will remain larger than the other, and no disease will be present, so that a horse is not necessarily spavined because one hock is larger than the other. A remark of Percivall may be quoted here: "Spavin, like splint and other transformations of soft, elastic tissues into bone, may be regarded as nature's means of fortification against more serious failures."

*Symptoms.*—Perhaps the most constant symptom is the tendency to stand on the toe with the heel elevated

5

when resting; the horse usually moves very stiff or lame for the first few minutes, which passes off in many instances after travelling a mile or so. If the inflammation has extended to the surface, there will be pain on pressure; later, there will probably be some enlargement. It will be found, however, that the symptoms of spavin vary considerably and depend on the situation of the diseased part, and the amount of destruction of normal tissues that is going on.

*Treatment.*—Have a high heeled shoe applied as soon as the nature of the disease is known, and, in my opinion, this is one of the most important requirements. If the lameness does not increase with exercise or perhaps light work, it is sometimes advisable to continue to work the horse every day. In fact, we frequently meet with cases that do not improve with rest, though sometimes persisted in for months at a time, but when turned out to pasture immediately do so, and soon become useful animals. This probably occurs in consequence of the process of bony union between the diseased bones becoming completed quicker when the bone is actively exercised. Other cases require long and absolute rest. The rule should be that all horses in which the lameness is increased by work, shall have absolute rest. The very best application as a remedy for spavin is HEARD'S AMERICAN EMBROCATION, well rubbed in three times a day for five minutes each time, until a blistering effect is produced. Then cease for five or six days after which repeat the treatment. This will cure any of the ordinary cases of spavin. If we have ulceration on the inside of the joint, and consequently great lameness, the horse should be fired and a blister composed of one part of biniodide of mercury and six parts of vaseline applied. A shoe with very high heel calks must not be omitted.

# RINGBONE.

WE shall consider an inflammation of the upper or lower pastern bones as ringbone, (see P. B. & R. D. in skeleton on page 1),or that frequent result of such inflammation, the formation of bony tumors on those bones. It is known as upper ringbone when the upper pastern bone is affected, and lower ringbone when the lower or short pastern bone is the seat of the disease. We not unfrequently have the joint between those two bones also involved, and in many instances the two bones become firmly united together (anclylosed) to form one bone.

*Causes.*—Inheritance is certainly one of the most common causes. Williams says : "I advise breeders of horses never to breed from a sire or dam having ringbones, unless their origin can be readily traced to some accidental cause." Age is an important consideration, as it usually occurs in young horses that are very early put to hard work. The shape of the limbs also has an important bearing on the frequency of the disease, horses with very upright pasterns being especially liable to it. This is what we might expect, for concussion to the bones will be much more severe in upright limbs than when they are more obliquely placed. Concussion and straining, and sometimes external blows are the active causes of this disease.

*Symptoms.*—The horse attempts to give the leg an oblique or slanting position, and therefore travels on his heels, except when the disease is located on the outside or back of the pastern of the hind leg, when he will put the toe down first, or what is known as "travel on the toe," if the disease is active there will be very severe lameness and some pain on pressure around the pastern. After some little time, depending on the extent and severity of the inflammation, there will be found an enlargement at some point on the pastern, sometimes involving the two pastern bones, as seen in fig 25.

*Treatment.*—If there is increased heat in the part and tenderness on pressure use hot water fomentations for half an hour, after which apply HEARD's AMERICAN EMBROCATION, rubbing it in well; then saturate a bandage with the same agent, and wrap it around the pastern. This should be repeated twice daily until the skin is well roughened. Then stop the active treatment for a few days, when if the horse is still lame, the treatment should be repeated. Unless the disease is situated on the outside or back of the hind pastern, there should be applied a bar shoe, made very thin at the heels. If the disease is located on the hind limb at the outside and back a high-heeled shoe must be used; in fact, it may be laid down as a rule that whenever a horse rests by standing on the toe or with the heels raised off the ground, a shoe raised at the heels should be applied. If the lameness still persists after repeated trials with the EMBROCATION, it will be absolute proof that the joint is involved, and the horse should be fired and blistered as soon as possible

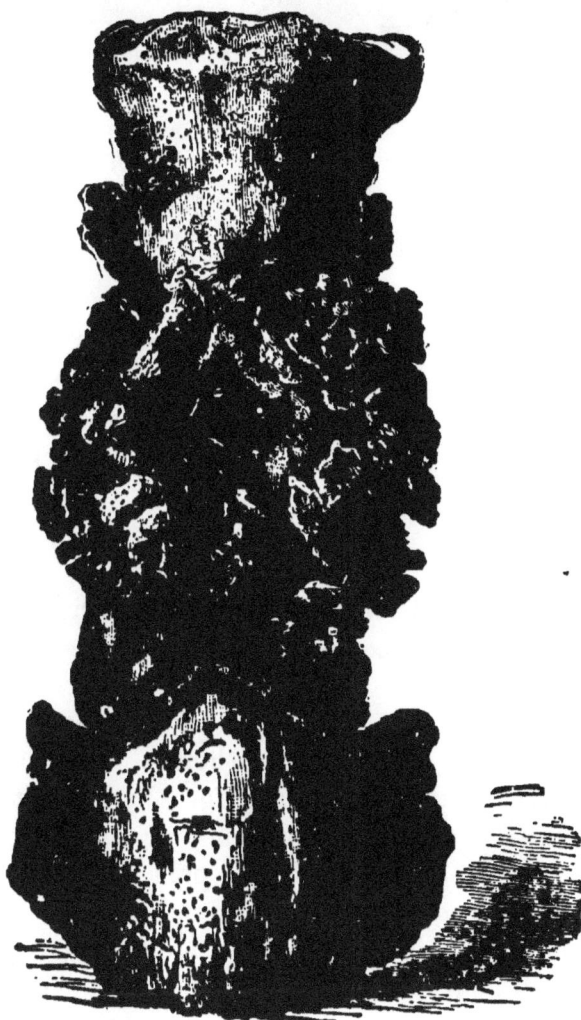

Fig. 25. Showing effects of ringbone on the bones of pastern

Fig. 26.   Coffin bone showing side-bones at A.A.

# SIDEBONE—OSSIFICATION OF THE LATERAL CARTILAGES.

At A.A. in Fig. 26 is seen the side bones which have taken the place of the cartilages; at B. is situated the coffin bone.

This is a disease that usually attacks the fore feet, and most frequently in heavy made horses. The change that takes place in the lateral cartilages, by which they are transformed into bone, is commonly a very slow one; in fact, takes place so slowly that no pain is felt and no lameness is noticed during development. At times however, the change is more rapid and inflammatory in its nature, and lameness becomes quite pronounced.

*Causes*—Inheritance, hard work at a very early age (principally drawing heavy weights), and active concussion.

*Symptoms.*—The sure sign is that the usually elastic

cartilages over the heels of the hoof become unyielding, stiff and hard. In a horse with normal lateral cartilages, slight thumb pressure over the quarters of the hoof will cause the cartilage to bend inward toward the pastern bone; but if sidebones have formed, there will be no give to the part when pressed with considerable force.

*Treatmemt.*—Rest, bar shoe, the firing iron and blisters.

## FRACTURES.

In fractures of bone we have a portion broken off so that it is not closely continuous with the remainder. We may have the bone simply cracked through without displacement, or it may be displaced to a considerable degree from its proper location. The bone may be broken straight across, or the crack may be in an oblique direction. The outside projecting points on bones are frequently the subject of fracture, as for example, the breaking off of a portion of the point of the ilium. (13 in figure of skeleton). In this article only the most common kinds of fracture will be treated on.

## FRACTURE OF LOWER JAW.

*Causes.*—Perhaps this bone is more frequently the subject of fracture than any other. It is usually caused by the bit bruising the gum and covering of the bone, so that a portion of it dies for want of nutrition, and a breaking off of the dead portion is the result. It is sometimes caused by the curb-chain bruising the covering of the bone under the jaw, and the break will take place the same as described above. It is sometimes caused by blows or external violence of some kind, when it is usually of a more serious nature.

*Symptoms.*—If from the bit, there will be swelling of the gums and considerable tenderness ; if the animal is worked, the saliva will dribble from the mouth in large quantities ; some horses get exceedingly ill-tempered when suffering from this accident, and will continually fight the bit if driven. Others again will not drive up to the bit and are apt to be made balky. If the curb-chain is the cause of the fracture, there will be swelling of the soft structures under the jaw, and very great tenderness. Abscesses sometimes form in this location as a result of the fractures, and we sometimes have formed a great deal of pus. When the fractured piece of bone has separated from the main bone, we will shortly get small portions of dead bone working toward the surface, so that when examining horses with sore mouths I quite frequently find sharp pieces of bone protruding through the open sore in the gum. This can be easily felt when the finger is run back over the gum. There is often a very bad smell from the horse's mouth. The horse will sometimes refuse to eat solid food, or only in small quantities, when the subject of this accident.

*Treatment.*—If it is caused by the bit, and the gum is simply swollen, the following lotion should be used 3 or 4 times a day: powdered borax, 1 ounce ; dissolved in half a pint of water, then add ½ an ounce of laudanum. If there is a wound in the gum and a bad odor from the breath, the finger should be inserted in the wound, and if any pieces of bone are felt, they should be removed. This can often be done by simply catching the piece with the finger and thumb and pulling it out ; at other times it seems to be wedged in tightly, and it will be necessary to use some kind of instrument to get it out. A forceps or a stout nail will usually answer the purpose. The nail should be run down alongside the loose piece of bone and the dead portion lifted out. Be sure that you have removed all the pieces, for if not, the wound will not heal. This should be followed by the application twice a day, with a small sponge or piece of cloth, of a little of HEARD's HEALING LOTION, and if the fractured pieces of bone are all removed, the wound will heal very quickly. The bit should not be used, as the pressure will retard healing and cause the animal great suffering. Feed soft food. If the swelling is on the outside of the jaw, and is caused by the curb-strap or curb-chain, it will be well to apply a poultice of linseed meal, to which has been added about a tablespoonful of HEARD's AMERICAN EMBROCATION, and repeat twice a day. If there are any wounds from ruptured abscesses, they should be dressed with HEARD's HEALING LOTION twice a day. If possible insert the little finger or a wooden probe to see if there is any loose bone. You may be sure that there is a fractured piece of bone if there is any very bad smell from the wound. This must be removed before the outside wound will heal. The curb-chain or strap must not be used until the wounds have healed.

# FRACTURE OF THE ILIUM.

The illeum is seen at 13 in the figure of the skeleton. Next to fracture of the lower jaw this is the most frequent. Sometimes a large part of the projecting spine is broken off from the main bone ; at other times only a small piece is chipped off.

*Causes.*—External violence, such as catching the projecting point of the illeum in passing through narrow doorways, falls on hard substances, etc.

*Symptoms.*—If seen soon after the accident, the side on which the fracture has taken place will look flatter than the other side. The fractured piece will have been drawn forward and inward by the contraction of the abdominal muscles, so that there will be quite a space between the two fractured ends ; consequently there will be no grating sound heard when the parts are moved about. The animal will be slightly lame at first which soon increases and may become quite pronounced. There will probably be some pain on pressing on the illeum. If a string is passed from the middle of the spine to the point of the illeum that has not been injured and the exact measurement taken, it will be found to be longer than the measurement when taken on the side of the fracture.

*Treatment.*—Rest, and the application three times a day of HEARD'S AMERICAN EMBROCATION, well rubbed in. This to be continued until the quarter is well blistered. With this treatment these cases always become useful for slow work and often make good driving horses. To a close observer, however—no matter how small a

piece of bone is chipped off—the subjects of this injury
will always seem to move slightly to one side behind,
and have a perceptible limp.    The best method to detect an old fracture of the illeum is to stand exactly behind the horse and look over the quarters, when any
uneveness of the two sides will be immediately detected.

# FRACTURES OF THE BONES OF THE LIMBS.

*Causes.*—Falls, kicks and blows of various kinds.
Sometimes the bones themselves are in an extra brittle
condition, when they are easily broken.    An illustration of this occurred in my own practice.    A horse was
being driven along the street when he suddenly gave
way on one side, and, trying to save himself on the opposite side, suddenly fell down flat and was unable to
rise.    On examination, both thigh bones were found to
be fractured at about the same place on each bone.
When the bones were examined, they were found to
contain an excess of earthy matter, and a very small
amount of animal matter, which latter makes the bones
tough and not so easily broken.    In the bones of old
animals the earthy matter is in much greater proportion
than in those of young animals.    The bones of old animals are therefore more easily broken.

*Symptoms.*—We may have fracture of the bones of
the limbs without displacement, the bone being simply
cracked across, or partly through, and not moved out
of its place.    In some of those cases there is not much
lameness.    The horse may even work again in a few
days, to be followed perhaps by a sudden fall while be-

ing driven, in consequence of displacement. A case
which occured in my own practice will be interesting
here. A horse was kicked in the thigh and the skin
somewhat bruised and a little swollen. When found in
the morning, there was scarcely any lameness and in a
few days he trotted out sound. He was driven two
days, apparently all right, when the owner concluded
to send him out to pasture for two months. About ten
days after going to pasture he was found in a good
roomy box stall with a fracture of the thigh bone at
about the point where the kick had been received. The
displacement probably took place when the horse attempted to get up in the stall, and although the kick
had been received three weeks previously, there was
no marked symptom of fracture being present. In most
cases, however, displacement takes place immediately,
and the limb can be swung around in unusual directions. When moved, a grating sound will be heard,
and a grating motion felt. The animal will be unable
to bear any weight on the fractured limb.

*Treatment.*—If the horse is of a nervous temperament
he had better be destroyed, as it is necessary for the
limb to have absolute rest to secure union of the fractured bones. If the bones are protruding through the
soft structures or through the skin, he had better be
destroyed. If he is a young, quiet, docile animal, and
the bone is simply broken in two pieces, the chances of
recovery are fairly good, and the following rules are to
be observed : If it is any part of the fore-limb, from the
elbow to the foot, that is fractured, the bones should be
brought together in the natural position. A padding
of oakum should now be placed around the limb. The
next thing to do is to place two or more thin pieces of

wood, such as lath, in front and behind the limb, so
that there will be no motion allowed.    Now apply a
bandage that has been soaked in starch and plaster of
paris.    Begin to apply the bandage from the bottom of
the limb and continue it upward until the whole limb
is well wrapped.  There need be no fear of the bandage
looking clumsy, by having too much bandage material.
The animal should now be placed in slings as seen in
figure 27.    The splints and bandages should be allowed

Fig. 27.   Horse placed in slings.

to remain on for six to eight weeks, unless the animal
seems to have excessive pain in the leg, with swelling

of the soft parts above the bandages, when suppuration under the bandage may be suspected. The bandage must now be removed, and if no abscesses or ulcers are discovered, it should be immediately reapplied. After the fracture has become reunited, there will still be some swelling around the bone at that part. This will be gradually removed by applying a small quantity of HEARD'S AMERICAN EMBROCATION every other day.

## LUMP-JAW IN CATTLE—BIG HEAD—
## ACTINOMYCOSIS.

THIS is a disease known to scientists as Actinomycosis and is found in man, cattle, and swine.

*Causes.*—It is in all cases caused by the growth of a plant fungus that takes for its sustenance, first the soft structures of the mouth and then the bone, spreading in this manner until, in old cases, a considerable portion of the jaw is destroyed. It is asserted by some authors that the fungus or its spores are sometimes found on the fodder that the cattle feed on. Especially is this true of barley and some other cereals. The spores of this fungus may be taken into the stomach and intestines, and the disease has been found in many of the internal organs as well as the jaw.

*Spmptoms.*—There will first be noticed a swelling at some part of the jaw. This slowly increases in size until it bursts, leaving an open sore which will not heal and prevents the animal from thriving or eating solid food. The disease is fatal in all cases that are allowed to run their course to the end.

An interesting subject to the breeder is as to what shall be done with the carcass of an animal that has been slaughtered while suffering from this disease. Shall it be destroyed? or can it be used for food? These are important questions, and the common opinion of the eminent authors who have written on this disease is that all the body, except those parts in immediate proximity to the diseased tissues, are eatable. An animal affected with this disease should be immediately removed from a herd, as it is liable to spread it among the other cattle. Fig. 28 shows a small piece of a cow's tongue, magnified 250 times (diameters). The fungus is seen at a and b.

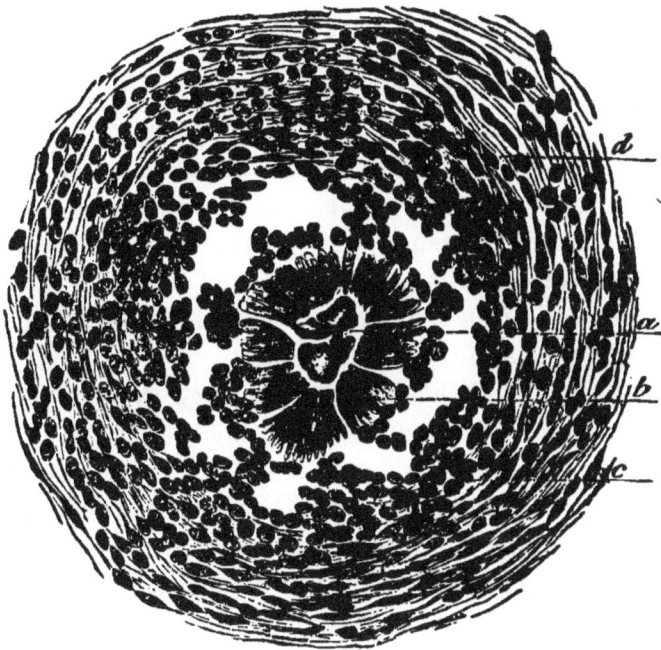

Fig. 28. Microscopic section of tongue showing fungus of actinomycosis at a and b.

# DISEASES OF JOINTS.

## INFLAMMATION OF JOINTS (ARTHRITIS).

*Causes.*—This is usually caused by external violence, but may be a result of concussion in traveling at a fast pace, or may be caused by sprains. We may also have a rheumatic inflammation affecting the joints. When an animal is the subject of kicks or bruises on the limbs, the important point is to note whether the injury is in close proximity to a joint or not, because an injury to a joint that has considerable motion is always a serious affair, and will require more than ordinary care in the management.

*Symptoms*—If we have inflammation of any of the joints of the limbs, there will be lameness, and the degree of lameness will depend on the severity of the inflammation. There will also be swelling and considerable pain on pressure. If severe, there will be fever and loss of appetite, the frequency of the pulse increased, and often a tendency to suddenly jerk the limb upward, as if suffering great pain.

*Treatment.*—If the lameness is very great, fomentations of hot water should be used, 3 or 4 times a day, immediately followed by the application of a small quantity of HEARD's AMERICAN EMBROCATION. If possible, apply a hot poultice of linseed meal, to which has been added about a tablespoonful of HEARD's AMERICAN EMBROCATION, this to be renewed twice a day. The following pill should be administered : Aloes 5 drams, ginger 1 dram and syrup sufficient to give it the proper consistency. Perfect rest is absolutely necessary, and it will be found beneficial to place the horse in slings, as seen in Fig. 27. If the injury is slight and the animal is not very lame, it is still necessary to enjoin absolute rest, and not to allow any motion until recovery is complete. Fomentations of hot water and poultices with the application twice a day of HEARD's AMERICAN EMBROCATION will soon effect a cure in light cases.

A case which occured in my own practice will serve to illustrate the necessity for rest. A horse fell in the street and injured a fore-limb. I saw him two days afterward and found him quite stiff and swollen in the vicinity of the elbow joint. I ordered fomentations and EMBROCATION, and at the end of five or six days the animal trotted out sound. The owner being anxious to work the animal—against my protest—took him out and worked him two days, and the third morning sent for me. I found the horse much lamer than when I saw him first. The treatment was repeated, but at the end of three days was followed by the soft parts breaking away and leaving an open elbow joint. I immediately told the owner the case was hopeless. He then sent for a professor, who now presides over a veterinary college in New York, who said that the horse could be cured if he was removed to his hospital. This was done and the result was that at the end of three weeks the horse was shot. We may always be sure that an inflammation of a joint which has much motion, is a very serious matter.

6

# OPEN JOINT.

*Causes.*—This is usually caused by external violence and frequently by sharp, pointed substances penetrating the parts covering a joint. It sometimes follows an inflammation in a joint where there has been no puncture.

*Symptoms*—Intense pain ; if in the limbs, there will be excessive lameness ; temperature raised, appetite poor, and a discharge of a yellowish semi-oily fluid from the wound ; this usually escapes in large quantity. The animal frequently jerks the leg up, often refusing to place the foot to the ground at all. If it is a joint of great motion, as the elbow, knee, or fetlock in the fore-limb, or the stifle, hock, or fetlock of the hind limb, there will be very little hope of cure. The only way a cure could result would be by having a bony union effected between the two bones forming the joint. This would result in a stiff, immovable joint, which would leave the horse comparatively useless for work. If the pastern joints are the ones affected, a stiff joint will not render the animal useless, as they are joints of quite limited motion.

*Treatment.*—If the joints of great motion are affected, the animal had better be destroyed. If the joints of limited motion are the ones affected, keep the outside wound well open, and thus allow the full discharge of pus and other inflammatory products from the joint. The external opening may be kept open by inserting a red hot iron. This should be done as often as required. While there is any discharge, the outside should be kept open, for it is a rule that all wounds have to heal from the bottom. Hot fomentations and poultices are beneficial. The horse should be kept in slings.

# DISLOCATIONS.

THE displacement of bones from their normal position in a joint is comparatively a rare condition in the horse and ox. But in dogs it is a frequent accident. Although it is possible to have dislocations of many of the joints in horses, there is only one that is common enough to require notice in this short work.

**DISLOCATION OF THE PATELLA, OR DIS-PLACEMENT OF ONE OF THE BONES OF THE STIFLE.**—This bone is seen at x in the figure of the skeleton on page 1, and is the analogue of the knee-pan in man. This is very frequently the seat of dislocation, the patella becoming displaced by slipping out over the prominence on the femur.

*Causes.*—The patella is kept in position by three ligaments holding it down, one on each side and the other in the middle. The eminence on the femur also assists to keep it in its proper place. Above, it is held in place by some of the muscles on the front of the thigh. Now, I am of the opinion that the under ligaments sometimes become unduly stretched, so that they fail to hold the bone down far enough. It may be said that they become too long, and the contractions of the muscles above will pull the bone from its normal position. At any rate we know that some horses while even standing in the stall, if made to move, will displace the patella, and when made to change position again, the

bone will slip into its place again. It was only last
summer that I had under treatment a horse that would
frequently displace first one patella and then the other,
and if made to change position, they would slip into
place again. The treatment continued for about three
months, the dislocations becoming gradually less frequent. When the horse first came under my treatment
the accident occurred several times a day. Only on
one occasion did I have to use force to replace the dislocated bone.

*Symptoms.*—The horse will be found perfectly stiff
in the dislocated limb which will be dragged along if
made to walk, it being impossible for the animal to
bring the limb even with its fellow, until the dislocation
is reduced. The position of the limb is seen in figure
29.

Fig. 29.  Position of hind limb with dislocated stifle.

*Treatment.*—Reduction can frequently be effected by
simply taking hold of the bone and pushing it inward
and downward, when it slips into its place. At other
times it will be necessary to attach a rope to the pas-

tern.  An attendant standing in front pulls the limb forward while the operator grasps the patella and pushes it inward and downward, using as much force as is necessary to accomplish the reduction, as seen in

Fig. 30.   Method of reducing dislocation of stifle.

figure 30.   It is well to bear in mind that bringing the limb forward relaxes the muscles that are attached to the upper part of the bone, and the operator will not have as much force to overcome as he would if the muscles were kept tense by the limb being held back. Having reduced the dislocation, the animal can walk as well as ever ; if lame it will be only in a slight degree; but remember that the ligaments are now stretched, and that the bone will easily slip out of place again. Steps should be taken to retain the bone in place and cause a shortening of the ligaments.   This can be effected by placing the horse in a single stall ; tie him short enough to prevent his lying down ; place a collar on the neck and attach a rope about 10 to 12 feet long to a leather strap, buckling the strap around the pastern of the limb that was dislocated. Now bring the rope forward between the fore-legs and tie it around the collar. If the limb is a little in advance of its fellow, it will prevent another dislocation. Now apply HEARD'S AMERICAN EMBROCATION around the stifle, rubbing it in well three times a day until a good blistering effect is produced. Let the horse stay in this position for two or three weeks. Although I have treated scores of cases by this method, I have never known it to fail in effecting a perfect cure.

## CONTAGIOUS DISEASES.

It is not my intention to describe all the contagious diseases to which the domestic animals are liable, as that would take up more space than is contained in this small volume. I shall therefore only notice two or three of the most common and give such advice on management and prevention of the spread of the diseases as any stock owner can easily understand and carry out. Some of the common contagious diseases that I shall omit to describe are spinal-meningitis in the horse, pleuro-pneumonia in cattle and cholera in hogs.

## INFLUENZA, STRANGLES, DISTEMPER, PINK-EYE, EPIZOOTIC, COLT-ILL, HORSE-AIL.

It is not certain whether all the various forms of disease known by the above names are caused by the same species of germ or not. Judging from clinical experience, it would seem that they are only varying forms of one specific disease. It is commonly said that one attack of distemper will prevent a second attack; but this is certainly erroneous, as I have repeatedly seen the same horse suffer from two or three attacks of this disease.

*Causes.*—Young horses are much more liable to this disease than horses of mature age, although we quite frequently see it in aged horses. Horses kept in small and badly ventilated stables are more liable to the disease than those kept in large, well ventilated stables. Horses that are hard worked, especially while young, will suffer more severely than those that are kept in a strong, hardy condition. Change of food will also act as a cause, as when horses are shipped from the farm—where they are fed on some kind of soft food—to large cities, where they are immediately placed on a diet of dry food and usually in much larger quantity than they received while in the country. Undue exposure to inclement weather is frequently a cause. It will be noticed that the various causes above enumerated all act by depressing the vitality of the horse, thus making him an easy prey to disease.

It is now perfectly well understood that all contagious diseases are caused by some living organism gaining access to the interior of the body, and there setting up a destructive action of some kind. Influenza in horses is no exception to this rule, although the particular germ which causes it has not been isolated and experimented with to the extent that it can be easily identified. Several observers have described germs which they assert to be the cause of influenza in horses, but as yet there is no particular germ accepted by all experimenters as the cause of this disease. The germs, however, are certainly the active cause of all forms of distemper.

How do the germs cause the disease? Most likely they act in several ways, each setting up a special form of the disease. We will first describe the most common form, where we have about the following:

*Symptoms.*—The animal is first noticed to lag more than usual at work; then a loss of appetite, or, if a mild case, a partial loss of appetite. Often we have cough ; the glands under the jaw are often somewhat swollen at this time, and the limbs soon begin to swell about the fetlocks. If the temperature of the animal be taken, it will be found from 2 to 7 degrees above the normal ($98\frac{1}{2}°$ Fah.). The pulse will be much quickened. This is usually followed in a day or two by a flow of mucus from the nostrils, the cough increasing in frequency. If properly treated, these symptoms usually begin to abate about the fifth day, gradually decreasing until the appetite is fully restored and the animal is again fit for work.

We will now describe a more severe case. To the above symptoms may be added a very dull appearance, the head straightened out and held down as if asleep, great disinclination to move, and often the subject of chills after drinking water. This severe form is often complicated with pneumonia, the lung trouble being a result of the general interference with the action of the heart and blood circulation. We always have a high temperature in this form, often reaching 106° Fah. on the second or third day after the attack. If the lungs are involved, the breathing will be quickened and the animal will rarely lie down. The appetite is totally lost, and by placing the finger in the mouth it will be usually found hot. In about two days the animal will frequently have a rambling gait, especially of the hind limbs ; seems to have very little control over the muscles of the hind limbs. If the throat is very sore and much swollen inside, there will be a considerable dribbling of saliva from the mouth, and when the patient attempts to drink, the water taken in by the mouth will frequently escape through the nostrils.

This is a very severe form, and unless great care is used in the management, it will often prove fatal.

The "Strangles" form is seen where we have, besides many of the above-mentioned symptoms, the formation of abscesses under the jaw. We sometimes have very large swellings develop in this location in a day or two. At first they are usually hard and very tender, and here we may have a total disinclination for food—at most only a little hay—on account of the abscesses interfering with the muscles used in the act of swallowing. The saliva will usually dribble from the mouth quite profusely. Sometimes these abscesses are deep enough in the throat to prevent breathing with any considerable freedom; in fact,in severe cases the obstruction to breathing causes very great distress. and the loud sound made by the air passing through the constricted passage in the throat can be heard for a considerable distance, In such cases there are frequent spasms of coughing, which sometimes last for two or three minutes. This form of the disease is not nearly so dangerous to life as the form last described, but when improperly treated, frequently leaves the animal a roarer. This is especially the case in race horses.

One other form of influenza may be considered typical, viz., pink-eye. Here the most prominent symptom is a swelling of the eyelids and a great tendency to have swellings of the limbs, due to an effusion under the skin. The eyelids are often seen swollen so large as to be entirely closed, and the attack seems to come on very suddenly. The general symptoms are much the same as those described for the mild form of distemper. We have sore throat, high temperature, quick pulse, loss of appetite and a rapid loss of strength.

I will notice still another form, which, however, is usually of a very mild character, a form which every veterinary surgeon of experience will recognize when I mention measles. This form is characterized by eruptions which frequently cover large surfaces of the body; in fact, the subjects of this disease are frequently covered all over with pimples. There is usually a slight rise of temperature, with sore throat, and often a loss of appetite. I have known horses affected with this form to loose all the hair on the body and be completely bare until the new coat came out; but this is an unusual occurrence. Except that the course of the disease is usually limited to a few days, it is very suggestive of measles in man.

As to grouping all the above described forms of disease under the general term "Influenza," it is suggested by the fact that we often find all the various forms exhibited in one large stable during a single outbreak. Some horses will be affected with one form, others with other forms of the disease at the same time. It is usually more prevalent in the spring and fall of the year, but an outbreak may occur at any time. The great epidemic of 1872 occurred in November and December, and spread over the entire country in about five or six weeks, in this respect, as in some others, resembling the great grip epidemic in man of 1891.

*Treatment.*—If the simple form first described be present very little medicinal treatment will be required. The most important will be to apply HEARD's AMERICAN EMBROCATION to the neck, under the throat, and down along under the windpipe three times a day, rubbing it well in each time. See that the surface of the body is kept warm by sufficient clothing and bandages on the legs. The nasal discharges should be frequently sponged off by soaking a sponge in a solution of Boracic Acid— two teaspoonfuls in a pint of water. Restrict the diet to a small quantity of bran, or grass, if in season. Two quarts of bran made into a mash twice a day will be quite sufficient. If the animal improves, give it a quart of oats three times a day as soon as the fever has disappeared. Absolute rest is required until recovery is complete.

If this simple form runs into the severe second described form, we must take more energetic measures about as follows : Place enough clothing on the horse to keep him comfortably warm; bandage the legs; keep the food away entirely for 36 or 48 hours; have water constantly in front of the patient; rub the neck with HEARD'S AMERICAN EMBROCATION three times a day until the skin is roughened, and administer, three times a day the following made into a ball: Quinine, one dram, nitrate of potash, six drams, camphor, one dram, with sufficient syrup to give it the proper consistency. This to be continued daily until the serious symptoms disappear or abate. Dosing a horse with half an ounce of quinine a day would have been thought quite extravagant some years ago, but with the price at 25 cents an ounce, as at present, it is not an expensive treatment and will be found very effective. If this form is complicated with pneumonia, we must treat the same as is laid down for that disease on pages 110 and 111.

*Treatment for Strangles Form.*—Here the same general treatment as regards clothing and diet is indicated; also the administration of the quinine ball as above described. To the enlargements or abscesses under the jaw, apply poultices of linseed meal or turnips, to which

has been added about a tablespoonful of HEARD'S AMERI-
CAN EMBROCATION, to be changed twice a day. As soon as
they become soft, the abscesses should be opened by
puncturing with a sharp knife, and the finger should
be inserted into the wound, and all the pus squeezed
out of it. Now dress the wound by soaking a little pad
of oakum or a small clean rag with HEARD'S HEALING
LOTION—which is a perfect antiseptic—and insert it into
the wound, allowing it to remain there 12 hours. This
method of dressing should be repeated twice a day
until the wound becomes so small as to prevent the insertion of the plug, when the sore is to be simply saturated with the Lotion to prevent the formation of
proud flesh. If the breathing is much obstructed and
suffocation is threatened, the operation shown in Fig. 31
must be performed. This is exceedingly simple, and

Fig. 31. Performing the operation of Tracheaotomy at night.

no possible injury can be done if the following directions are strictly followed out : with a sharp knife make
a slit through the skin about 4 inches long ; now cut
and divide the soft structures until the windpipe comes
into view ; then push the knife through and cut out a
portion of two of the rings of the windpipe, making a
circular opening about three-fourths of an inch in diameter.  Place a tube in this opening ; but this is not
immediately necessary.  To prevent suffocation the operation can be performed by any one, the opening being
simply allowed to remain without a tube until one can
be obtained.  There are various kinds of tubes sold by
instrument makers for this purpose, but as it is my intention to enable the ordinary breeder or stock owner
to save the life of his stock in cases of emergency, I
will describe a very simple tube which will answer the
purpose, and which can be obtained in a few minutes
at any tinshop.  The tube shown in Fig. 32 should be

Fig. 32.   Simple tube for insertion in trachea.

about five-eighths of an inch in diameter, the length of
tube to be about 3 inches.  The two rims seen in
the figure are for the purpose of making a hold to
attach a piece of tape to tie around the neck.  This will
hold the tube in place and prevent it from slipping out
of the windpipe.  After the opening is made, relief will
be instantaneous.  The tube should be kept in several

days. The condition of the throat can be easily tested by simply placing the hand in front of the tube and compelling the horse to breathe through the nostrils again. As soon as the animal can breathe in this manner without showing any obstruction in the throat, the tube should be removed and the wound dressed twice a day with HEARD's HEALING LOTION. It will close up in a few days. It is sometimes necessary to allow the tube to remain in the trachea 2 or 3 weeks before the obstruction in the threat is entirely gone. If it gets stopped up with mucus, clear it out by scraping with a knife or piece of wood. Keep it clear. The diet should be restricted to about 4 quarts of milk with water in which oatmeal has been soaked for an hour or two. If the case is very protracted, a nice nutritious drink can be made by boiling, for 15 minutes, half a pint of linseed in a gallon of water, giving it as often as the patient will take it. After a day or two a small quantity of hay can be given, and a little bran mash will sometimes be relished. But it is much better to restrict the diet to very small proportions.

The treatment of the next described form, pink-eye, will have to be varied in only one particular, and that relates to the local management of the eye. It is important that the inflammation of the eye-lids and surrounding structures should be reduced as soon as possible, otherwise we are very likely to have a horse with defective sight, and consequently of greatly reduced value. To reduce this inflammation there should be inserted, with a syringe, about a teaspoonful of HEARD's LOTION for moon-eye, to be repeated 3 times a day. There is a great tendency in some horses to rub the lids against the side or front of the stall. Prevent this by turning the horse around in the stall, and tying him to two posts so that he cannot get a chance to rub the eyes. This is a very important part of the treatment. The diet, clothing and medicinal part of the treatment must be similar to that described for the treatment of the severe or second form of influenza. If there is any whiteness on any part of the front of the eye, the application of HEARD'S EYE LOTION should be continued until it has all disappeared.

In the treatment of the measles form a half dram of calomel should be added to the ball before described, the diet similarly restricted, with the application of a wash to the eruptions made by mixing one-fourth HEARD'S HEALING LOTION and three-fourths water. The nostrils should be kept as clean as possible by frequently wiping off the mucus with a wet sponge or cloth. The legs will frequently remain swollen for some time, but with exercise, after recovery, this will gradually disappear.

## TUBERCULOSIS.

THIS is an infectious disease common to many species of animals. Man is especially liable to it, in whom it is the disease known as consumption. Cows, hogs, and poultry are frequently the subjects of it. It is extremely rare in horses and dogs. Perhaps there is no disease which has attracted so much attention in the last decade as tuberculosis, yet we have not discovered any method by which it can be detected in all cases. On account of the difficulty of making a diagnosis, even by experts, I shall not attempt a detailed article on this disease, but simply refer to a few general facts relating to it that should be known by all owners of cattle. It is asserted by very competent observers that about 5 per cent. of the cattle in this country are affected with this disease. It is usually propagated by the germs being inhaled into the lungs with the air. It is also propagated by the germs being taken into the stomach with the food or water. In stables where a large number of cattle are kept it is very liable to spread if there is an infected cow in the herd. It is not nearly so fatal in cattle as in man. It is only when a considerable surface of the lungs or some other organ becomes involved that disease will be suspected. In the first stages of the disease not even an expert could detect it unless he should by chance find the germs in the discharge from the nostrils, that is, if there is such a discharge. The disease sometimes attacks the udder and teats of cows. The germs are then frequently found in the milk itself, rendering the latter dangerous to the health of calves or human beings that partake of it. The disease also occasionally attacks the joints of cows, causing them to ulcerate. The cow becomes emaciated and either dies or has to be killed. When the lungs become badly affected there will be a cough and often a discharge from the nostrils. If several cows should show these symptoms, tuberculosis will very probably be present, and an expert should be immediately called in to discover, if possible, how far the disease has progressed in the herd.

## TETANUS—LOCK-JAW.

THIS disease usually follows wounds, especially punctured wounds, and more especially the wound called,

7

" punctured foot," caused usually by "picking up nails." It is also caused by a germ. The reason that it so often follows punctured wounds is that the entrance of air into such wounds is prevented by the walls of the wound immediately closing. Investigations have recently discovered that the germ that causes lock-jaw cannot grow and propagate if exposed to the air ; but if plante I in suitable material and air be excluded, it thrives and is propagated with rapidity. This germ, when growing, generates an intensely active poisonous substance, having much the same properties as strychnine. This poison is absorbed into the system, and, acting on the central nervous system, causes the pecu-

Fig. 33. Appearance of horse suffering from Lockjaw

liar muscular contractions seen in lock-jaw. There is
not the least doubt that this is the true cause, as recent
experimentors have frequently cultivated the germs
out of the body and filtered them so as to separate them
from the poison, and then injected the poison into animals and produced true lock-jaw.

*Symptoms*—The first thing noticed usually is that the
animal moves rather stiffly ; this is soon followed by an
inability to take food, although the attempt to eat is
frequently made. The head is now straightened out,
and the animal will present the appearance as seen in
figure 33. If an attempt is made to turn the patient
around, it will be found that it is unable to bend the
body. The muscles will be very hard to the touch.
On account of the spasm of the muscles of the throat,
the saliva will dribble from the mouth in large quantities.

*Treatment*—The principles of treatment that are here
indicated are to attempt to stop the growth of the
germs, and consequently the formation of the poison,
by allowing free access of air to all parts of the wound ;
and to drive the poison out of the body ; also, if there
is much excitement, to give something to keep the animal quiet. First, then, the wound should be opened so
that every part of it is fully exposed to the air.

Second, give the following draught as soon as possible :
barbadoes aloes, $1\frac{1}{2}$ ounce, powdered, and well mixed
with a pint of warm water.

With this may be given a pint of linseed oil. It is
frequently very difficult to administer medicine in any
form by the mouth. I have occasionally caused purging by injecting a fluid solution of aloes into the rectum. The animal should be given no food for at least

two days, and must be kept in a quiet, dark stall. Allow the patient as much water as it will take. Dress the wound twice a day with HEARD's HEALING LOTION, and if the patient seems to suffer much pain, administer a bottle of HEARD's MAGIC MIXTURE. If the animal inproves, give it soft food, oat-meal drinks, grass if in season and linseed-tea. I have had success in a number of cases of this fatal disease by the above mentioned treatment.

## GLANDERS—FARÇY.

THIS is one of the most dangerous and fatal diseases to which the horse is liable.

*Causes.*—An animal that has suffered from any debilitating disease,such, as for example, distemper, is more liable to contract the disease than one in perfect health. Abrasions or sores on the gums or lips will afford easy access to the germs. In a large number of cases this disease is taken at a public watering trough; sometimes in blacksmith shops, sometimes in livery stables, where strange horses are frequently put up for a short time to feed, etc. In every case we may be certain that it is taken from a pre-existing case of glanders, although the animal from which the disease is taken may not have been in contact with the inoculated animal at all, nor in fact been at the place of infection for months. It has been asserted that the germs of glanders will live for many months when deposited on boards or other materials which enter into the composition of stables. The germs may be mixed with mucus and be deposited on the boards in front of a stall; another animal rubbing its lips on the boards is liable to take the disease.

The growth of the germs of glanders is very slow when compared with many others, but they are extremely tenacious of life.

*Symptoms*—The germs having gained access to the body, they usually cause small tubercles to be formed in the lungs and air-passages.

When these tubercles break down there will be a discharge of pus and mucus from the nostrils. The first thing noticed usually is that there is a slight discharge from one nostril with a swelling under the lower jaw about the size of a walnut, rather loosely connected, and not very tender to the touch. This may continue about the same size for a long time, and as there are no constitutional symptoms, no dangerous disease is suspected. These are the really dangerous cases and the ones from which nine-tenths of the cases are propagated. The owner will usually say the horse has had distemper and hasn't got rid of it yet : that he eats all right and works well, in fact that there is virtually nothing the matter with his horse. I have known horses to keep fat and work well for a year or two with glanders in this mild and evidently local form. Always beware of a horse that has a movable swelling under the jaw and a discharge from the nostril on the corresponding side. After a while the disease assumes a more severe form ; there is a discharge from both nostrils ; the horse gets thin, cannot work with comfort any more, and we will be liable to notice ulcers on the mucus membrane of the nostrils which makes the diagnosis certain. Swellings may now appear on the surface of the body, which soon break and discharge pus and the resulting sores refuse to heal. .In a short time the animal will die.

At other times this disease will appear as farcy, a disease that is caused by the same germ, and in fact may have been taken from one of the above described cases. Farcy selects the skin for attack instead of the air-passages. It is indicated by little swellings on the skin, usually of the limbs. These soon break and discharge pus freely, and do not heal when treated as ordinary wounds. In a day or two the limbs are seen to swell considerably, and after two or three weeks they are a mass of sores.

*Prevention.*—It has usually been considered that all cases were fatal, but there are two or three facts that indicate an occasional recovery from a local infection. First, it is known that a glanderous ulcer in the nostril has healed. Second, there are a few pretty well authenticated cases of recovery from mild local attacks. Third, one of the most celebrated cases of recovery from glanders in man is that of Doctor Paquin, who was for a long time state veterinarian in one of the western states (Illinois, I think). Recovery, however, is exceedingly rare, and on account of the ease with which man is inoculated with this disease, all animals suffering from it should be immediately destroyed; in fact, nearly all of the states have laws to this effect. All suspected cases should be immediately isolated in a small pen or shed. I have known a single glandered horse to infect a herd of 40 when turned out to pasture. Therefore on no account should a horse that is even suspected to have glanders be turned out to pasture with other horses. And remember that the most dangerous case is the animal that is supposed to have very little the matter with it. All stables where horses with glanders have been kept should be thoroughly disinfected by washing all materials with which the horse may have been in contact with the following solution : corrosive sublimate, one ounce, dissolved in two gallons of water; everything to be washed with it twice a week for two weeks.

# DISEASES OF THE RESPIRATORY ORGANS.

## SORE THROAT.

It is a very common thing for horses to have an inflammation of some part of the throat. This part may be the tonsils, the palate or the larynx.

*Causes.*—Anything which predisposes an animal to take cold, as a debilitated state of the constitution, standing in drafts when perspiring, etc., are the most frequent causes of this trouble.

*Symptoms.*—There will often be slight fever in the first stages of this disease, with the pulse somewhat quickened. If the palate and tonsils are involved, there will be a great disinclination to take food, and if the mouth is opened it will be frequently found filled with saliva and mucus, which will dribble from it. Most horses suffering from this disorder will only take small quantities of water and often none at all for a day or two. Unless there is much swelling in the larynx, the breathing will be about normal. We may not have any cough, but if, by squeezing the upper part of the windpipe, a cough is forced, it will be very soft, as if the animal was afraid of hurting the already tender throat. In fact, the cough of sore throat is entirely characteristic. The horse sometimes holds the head somewhat straightened out. When the walls of the pharynx are

much swollen, if allowed to drink, the water will escape
by the nostrils as shown in Fig. 34.

Fig. 34: Representation of horse with sore throat and water escaping by the
nostrils while drinking.

*Treatment.*—Perfect rest; keep the animal warm
with plenty of surface clothing and bandages on the
legs, and rub HEARD's AMERICAN EMBROCATION on the throat
3 times a day. Place on the tongue with a spoon or
· flat piece of wood about a tablespoonful of the following paste : syrup, one-half pint ; chlorate of potash, 2
oz.; quinine, 2 drams ; oil of tar, 2 drams ; licorice
powder, 1 oz.; mix well. Give this 4 times a day. If
there is no other complication, this is not a dangerous
disease, but when badly treated the membranes of the
throat are sometimes permanently thickened, which is
occasionally the cause of roaring. The animal should
be allowed to properly recover from the disease before
being worked. When the appetite returns give HEARD's
QUININE CONDITION POWDERS 3 times a day as a tonic.

This will be found to produce the best possible condition in a very short space of time.

## ROARING.

THIS is a disease in which there is a loud sound produced by the air passing through a restricted tube during respiration, while the animal is undergoing great exertion. The constriction in the air passage is in the larynx, and in nearly all cases is due to a paralysis of the muscles on one side of the throat, which allows the muscles on the other side to draw the cartilages toward that side, and thus obstruct the passage of air through the windpipe. If the constriction is due to a thickening of the membranes of the throat, following a sore throat, there is good ground to hope for a recovery, and I have known many such cases; but when it is due to a paralysis of the muscles of one side of the throat, the case is a hopeless one, or at any rate is only curable by an operation, and then the chances are rather against a perfect cure. If the roaring follows a severe sore throat, it is always prudent to attempt a cure, and nothing will be so successful in bringing about that event as the administration of HEARD'S QUININE CONDITION POWDERS and the application 3 times a day of HEARD'S AMERICAN EMBROCATION, to be continued until the neck is slightly blistered. The animal to have gentle work.

# BRONCHITIS AND PNEUMONIA—LUNG FEVER —INFLAMMATION OF LUNGS.

ALTHOUGH for the professional veterinarian these two diseases should be studied apart and their individuality be kept distinct, the ordinary, every-day stock owner would be simply baffled should an attempt be made to distinguish them in a book of this description. For the convenience of the stockman I shall therefore consider them as one.

Bronchitis is an inflammation of the air tubes that ramify all through the lungs. Pneumonia is an inflammation of the structures of the lungs that lie outside of the bronchial tubes. But the two diseases very frequently exist together, and we find in most cases that the tubes and other structures of the lungs are all considerably involved in the diseased process.

*Causes.*—The most common is exposure to cold damp weather or drafts, especially when a long coat is saturated with perspiration. Consequently in the fall before horses are clipped, these diseases are always more or less common. It is sometimes caused by small, thread-like worms (filaria) in the windpipe or trachea. Calves and lambs are particularly subject to the disease from this cause in some localities. I have seen many cases of bronchitis follow careless drenchings (giving liquid medicines).

*Symptoms.*—Usually the first thing noticed is that the animal seems disinclined to move around freely or it may seem to lag in its work. The attendant will notice that it does not look as well as usual. After being allowed to stay quiet for half an hour in the stable, an examination should be made about as follows :

First—The number of respirations in the minute should be counted; if normal, they will probably number between 12 and 16 a minute. If the number is greater than 16 the lungs are probably diseased.

Second—By placing the ear to the chest behind the elbow, the heart will be heard beating against it. The beats should be counted; if normal, we shall find between 35 and 42 beats a minute. I will here mention that in some horses, especially if they are at all nervous, the heart beats will be much quickened when the ear is first placed to the chest, therefore the count should be made about two minutes after applying the ear to it. If the pulse numbers more than 42 to the minute, there will be some fever. The rise in temperature will be indicated pretty correctly by the pulse; and the experienced examiner can usually guess the temperature pretty accurately by noting the quality and number of the pulse or heart beats.

Third—Every stock owner should be supplied with a clinical thermometer for taking the temperature of animals that may be sick, and as they can be obtained for about $1.25, the price is no object when compared to the advantages to be derived from its use. Many a severe attack of disease would be cut in the bud if the owner of the animal could detect its beginning. A stock owner would soon learn the value of such an accurate aid as the clinical thermometer. If the temperature of the horse is above 102 degrees, Fah., there is disturbance enough in the system to make it obligatory on the owner to take some steps either to stop its further rise or to diminish it. It should be remembered

that exercise will always increase the number of respirations and pulse beats, and also the temperature,
so that this examination should be made at least a half
an hour or an hour after exercise or work. The temperature should be taken as follows : First shake the
mercury toward the bulb until it falls below 98½ degrees ; then insert the thermometer in the rectum,
bulb end first, till it is within a half-inch of the end,
which must be held between the finger and thumb.
Allow it to remain in the rectum 4 minutes and then
withdraw it. By noting the height of the mercury from
the bulb the number of degrees of temperature will be
accurately shown.

If the horse is suffering from bronchitis, there will
frequently be a cough from the first. But this is not
so noticeable in pneumonia without bronchitis. If
bronchitis predominates, it will be discovered by placing the ear in front of the chest just below the bottom
of the windpipe, and hearing a sound such as would be
produced by air passing through a fluid-like material.
This is a sure symptom of bronchitis. If pneumonia of
one lung only is present, it may be distinguished by
placing the ear over the two sides of the chest and
listening to the sounds ; there will be a difference between the sound on the diseased side and the normal
side. I shall forego an explanation of this difference,
as this work is written for the average stock owner and
not for the veterinarian. The importance of the discovery that the two lungs do not give the same sound
is very great—it determines the fact that one or the
other lung is diseased. I would only be causing the.
stock owner more difficulty in arriving at a correct conclusion, by complicating the description with matter

that would take up at least seven or eight pages of this
book if I were writing for the purpose of educating
veterinarians in the art of the discovery of pneumonia
in all or a small part of a lung. The ears of the animal
will usually be cold, but in some cases they are warmer
than usual. The appetite usually fails shortly after the
begining of the disease ; but some horses will eat considerable quantities of food, especially hay, until a few
hours before death ; and here lies the danger in these
cases. The stockman will not suspect a serious and
probably fatal disease to be present while the animal is
eating almost a normal amount of food. It cannot be too
strongly impressed upon the minds of stockmen that the
appetite is often fairly good even in cases of bronchitis
and pneumonia. In bronchitis we often have a discharge
of mucus from both nostrils, which becomes very profuse as the disease advances. In pneumonia we sometimes have a discharge, but here it has a pinkish appearance ; if the disease is severe it may somewhat
resemble blood. This is the so-called pneumonia exudate. There is only very slight pain in either bronchitis
or pneumonia ; in fact, almost none at all, except when
pleurisy is present with the pneumonia ; then the animal will usually grunt when made to turn around, and
will sometimes be heard to do so when standing in the
stall. A horse affected with these diseases will rarely
lie down, and when a horse that is in the habit of lying
down every night, is discovered to have remained
standing all night, he should be carefully watched to
see if it is repeated on the following night. This does
not apply to horses that rarely or never lay down at
night.

*Treatment.*—If discovered in time and remedial

measures are applied immediately, bronchitis and pneumonia are rarely fatal ; in fact, taking mild and severe cases together, I don't think the losses from these diseases in my practice will average more than 10 per cent. Of course if the animal is continued at work after the advent of the disease, the chances of recovery will be diminished. Rest should be ordered immediately, blankets should be put on to keep the surface of the body warm, bandages to the legs and a hood over the neck and head. Allow no food to be given for at least 24 hours, and then only one quart of bran mixed with a gallon of water. Allow the patient as much water as it will take. Give injections of warm water and soap twice a day, throwing about two quarts of water into the rectum at each injection. Allow good ventilation through the stable night and day; don't allow your stable to have a suffocating odor when you enter at night a few hours after all the horses are in and the doors closed. Administer a tablespoonful of the following paste four times a day : syrup, one-half pint ; quinine, 1 oz.; extract belladonna, one-half ounce ; to be well mixed and placed on the tongue with a spoon or a flat piece of wood. This treatment to be continued until an improvement takes place, when the paste may be given twice a day. Rub the sides of the chest and under the windpipe 4 times a day with HEARD'S AMERICAN EMBROCATION, as it gives instant relief to the tired muscles that are used in breathing, and gives them exactly the stimulation they require. After a day or two the patient should be given an oatmeal drink—made by soaking 2 lbs. of oatmeal in a pail of water for a couple of hours, the meal to be well rubbed up between the hands to squeeze out the most nutritious part of it. If the animal refuses to eat for 3 or 4 days, and the debility becomes pronounced, so that the horse sways from side to side when led out of the stall—a frequent thing in such cases—give about 4 quarts of milk a day until the appetite returns. Some patients will drink milk very readily, while others will steadily refuse it, and must be made to take it by being drenched with it out of a bottle or preferably a horn. A half a pint of whiskey, mixed with a pint of linseed tea and given 3 times a day by drenching, is often followed by an immediate change for the better. Keep this up while the intense debility lasts. As soon as the temperature of the patient is materially reduced, give a pint or two of oats twice or three times a day, allowing only a pound or two of hay a day while the temperature is high. If grass is in season, the patient should get from 2 to 6 pounds a day until the temperature is nearly normal. As soon as the appetite has been fairly well established, give HEARD'S CONDITION POWDERS as per directions on the can. It will be found that the strength will be very rapidly re-established. By carrying out the treatment here described, many animals that would die from this disease will make a good recovery. On no account must the animal be put to work before a complete recovery is established, and then the labor should be very light for some time.

## PLEURISY.

*Causes.*—Same as mentioned for bronchitis and pneumonia.

*Symptoms*—Mostly the same symptoms as described

for pneumonia, except that there is rarely any cough ; considerable pain, with frequent grunting, especially when the patient is made to turn. This is a much more dangerous disease because of the tendency to rapidly fill the chest with fluid, besides which it is frequently complicated with pneumonia. In fact, the inflammation of the pleura, if not checked, will very soon extend to the lung itself.

*Treatment.*—Pursue the same general principles of treatment laid down for bronchitis and pneumonia, with this variation : Drench with 4 oz. of linseed oil twice a day for two days. Apply a blister prepared by mixing one-quarter of a pound of mustard and one-third of a bottle of HEARD'S AMERICAN EMBROCATION to the sides of the chest, immediately behind the elbow. This should be well rubbed in for 5 minutes each side, and covered by a layer of paper. Don't omit the injections of warm water and soap. The diet should be regulated as directed for pneumonia ; also clothing, ventilation, and after treatment.

## BROKEN WIND—HEAVES.

THIS should not be confounded with the disease of the throat in which a loud noise is heard when the animal is forced to great exertion and known as "roaring."

In broken wind the abdominal muscles will be seen to heave strongly at the flanks when the animal is put to any considerable exertion. There is also a peculiar cough, which is quite distinctive, and this is often accompanied by an escape of flatus from the anus. There is nearly always a disordered condition of the digestive system, with an extraordinary appetite. As this is an incurable disease, the treatment can only be palliative. The most important point is to attend strictly to the diet. Give only small quantities of hay, not more than 6 lbs. a day, in two portions, for an ordinary horse. The administration of Dr. HEARD'S CONDITION POWDERS will cause a great improvement in the breathing, and if given steadily, many wind-broken horses will scarcely show the effects of the disease, when put to ordinary work. Its effect is also to greatly diminish the tendency to cough.

# DISEASES OF THE DIGESTIVE SYSTEM.

## CHOKING.

IT is quite common for the veterinarian to be called to see both horses and cows suffering from this accident.

*Causes.*—A constriction in the gullet is sometimes the cause of an obstruction to the passage of food into the stomach. Cows are often choked by having an ear of corn impacted in the gullet. Other obstructive agents are potatoes, turnips, beets, carrots, and sometimes hay; and I have seen several cases in horses that were caused by the impaction of oats and even grass.

*Symptoms.*—In the horse there is a very profuse flow of saliva and mucus from the mouth and nostrils; spasms of the muscles of the neck will occur every two or three minutes; the head will be drawn back, the neck arched and the animal will show great uneasiness as seen in

8

Fig. 35. A swelling will often be noticed on the left side of the neck in the grove, above the windpipe. The horse will often cry out with the pain when the spasms come on.

Fig. 35. Horse choked by an obstruction in the gullet.

*Treatment.*—If the choking is in that part cf the gullet forward of the chest, and the swelling caused by the obstruction can be seen and felt, an attempt must be made to dislodge it by moving it up and down with the hand, using great pressure if necessary. If this does not succeed, half a pint of linseed oil should be given as a drench. This lubricates the walls of the gullet and aids the passage of the obstruction. If after working in this manner for an hour or so the obstruction still remains, an attempt mnst be made to force it along by mechanical means. The instrument known as a probang is used by veterinarians, but as the average stockman would not be able to obtain its use readily and as in many cases time is precious, an instrument must be improvised. A stiff new rope, about three fourths of an inch in diameter will usually answer the purpose. Open the mouth and give the tongue to an assistant to hold, while the operator pushes the rope back over the tongue and through the pharynx to the gullet. Push it along until it reaches the obstruction; then turn it around by a rotary motion, good pressure being exerted at the same time. In this way the obstruction will usually be passed along the gullet and into the stomach. Withdraw the rope. There is no danger in this operation. Immediately after the obstruction is removed, the muscular spasms will cease, the animal will appear to be easy, and the pain have disappeared. The horse should now be fed on soft, sloppy food for about 24 hours to prevent a recurrence of the trouble.

## STOMACH STAGGERS, SLEEPY STAGGERS, OR GORGED STOMACH.

*Causes.*—Over feeding is the main cause—often the result of a horse being loose at night, gaining access to the grain bin, and gorging itself with food.

*Symptoms.*—As a result of decomposition or chemical change of the food in cases of this disease, there seems to be formed a product which has a very strong toxic effect on the nervous system. The animal will probably be found standing in the stall with the head drooping or perhaps with the forehead pushed up close to the wall in front of the stall. When asked to move in the usual way, it takes no notice of the request. If made to move over in the stall by pushing, it will probably stagger, and almost fall down. If you take it by the halter and push it back a step or two, it is likely to push forward again as soon as allowed freedom of the head. This condition is well represented in Fig. 36.

Fig. 36. Horse suffering from stomach staggers.

Horses remain in this condition for days at a time, but by appropriate treatment gradually regain their normal powers of motion. They sometimes get worse, delerium comes on, and they know nothing that is going on around them. They may become furiously mad or utterly paralyzed and helpless, which increases until death.

*Treatment.*—No time should be lost in getting the following medicines into the patient: one ounce of barbadoes aloes, powdered, and mixed with a quart of linseed oil, to be given as a drench. Injections of soap and warm water to be given every three hours. One oz. of

bicarbonate of soda, disolved in a pint of water, to be
given every four hours. If the bowels have not moved
briskly in 36 hours, give a pint of linseed oil, and repeat every 12 hours until they do so move. Give no
food, but allow as much water as the animal will take.

## FLATULENT OR WIND COLIC.

*Causes.*— In these cases we have great distention of
some of the abdominal viscera by gases that are formed
as a product of the decomposition of food. It may follow a hearty meal of grass, or even an ordinary feed of
oats.

*Symptoms.*— The animal usually shows the first
symptom of uneasiness by pawing; soon an attempt is
made to lie down; the belly soon begins to enlarge;
there is swelling at the flanks; perspiration becomes profuse; breathing becomes somewhat labored and difficult;
eructation of gases often takes place, when the stomach
is the seat of the disease, and food is often ejected from
the stomach through the mouth and nostrils in considerable quantities. I know that most authorities have asserted the impossibility of the latter proceeding, but I
have witnessed it in many cases, and in which recovery
subsequently took place. The distention is sometimes
so great that the animal in tumbling around ruptures
the stomach or perhaps an intestine. This allows
the escape of the contents of the viscera, and is sure to ·
end in death.

*Treatment.*— When first noticed give a bottle of
HEARD'S MAGIC COLIC MIXTURE, and begin to give the
soap and warm water injections every hour. Now apply some of HEARD'S AMERICAN EMBROCATION to the skin

under the abdomen, rubbing it well in.    This often
starts an immediate action . of the
bowels.    If not relieved in an hour
and a half, give another bottle of
the MAGIC MIXTURE, and apply the
EMBROCATION as before.  If at the end
of another hour the gases are not
freely escaping from the rectum, the
operation of tapping should be performed as follows : with a trochar,
seen in Figure 37, make a puncture
as seen in Figure 38, as deep as the

Fig. 37.  Trochar for puncturing intestine in flatulent colic.

Fig. 38.   Place to puncture in flatulent colic.

length of the trochar will allow. On withdrawing the trochar the gases will usually escape until the distention nearly or quite disappears. The animal immediately becomes easy, and will frequently remain so. If you do not succeed in striking the gases at A in the above figure, a puncture should be made at B, in the floor of the abdomen. I have never seen any bad result from this operation, except the occasional formation of an abscess on the side, which is easily cured. Give the patient no food for 24 hours, and then very sparingly. Allow plenty of water to drink.

## SPASMODIC COLIC.

*Causes.*—This form of colic is produced by spasm of a portion of the intestine. It is frequently caused by the animal taking too large a quantity of indigestible food, as the straw used for bedding, &c., or a large quantity of cold water when in a heated condition; also by exposure to cold rains, over-driving for long distances, etc.

*Symptoms.*—A notable fact is that the pains come and go at short intervals. The intermissions will vary in all cases, sometimes being only a minute or two ; at other times the animal may enjoy a rest of at least 15 minutes, when the pains will suddenly come again. The pain is rarely continuous in the early stage of the disease. The presence of pain is shown by the animal pawing, lying down, and in some cases kicking around violently, rolling over, etc. There is great uneasiness generally. As the disease progresses the intermissions between the pains become shorter and shorter until pain becomes continuous. The animal often makes frequent attempts to pass the urine, without passing much at any one time; from this fact the owner is usually convinced that it is suffering from disease of the kidneys, or bladder ; but although I have made examinations of the bladder in hundreds of such cases, I have never found it distended with urine in but one case, and in that case I drew off the water by the catheter. The animal continued to suffer pain for some time after, thus showing that the bladder distention was not the cause of the pains. In fact, about the only disease of the urinary organs that would cause such severe pains would be stone in the bladder, which is a very rare occurrence.

*Treatment.*—As soon as an animal is taken with this disease, give a bottle of DR. HEARD'S MAGIC COLIC MIXTURE, and the effect will really seem to be magical, for in many instances the pains cease in a few seconds and the animal becomes quiet. If the disease is allowed to progress for some time before the mixture is given, the medicine is not absorbed so readily, and the effect is not so sudden and well marked. Because the animal has become quiet, it is not always safe to treat it as if there had been nothing ailing it; but injections of warm water and soap should be given every hour until the bowels move freely. Allow a moderate quantity of water to drink, but no food for 12 hours, and then only a small quantity. If the pains return, or if they do not cease within an hour or so, give another bottle of the MAGIC MIXTURE, and apply very hot fomentations to the abdomen for 15 minutes, after which rub in a couple of tablespoonfuls of DR. HEARD'S AMERICAN EMBROCATION to prevent the animal being chilled. This will cure all cases of ordinary colic. When it fails, if you will take the trouble to make a post-mortem examination, you will most likely find one of the following conditions; Gut-tie, intussusception (the doubling of a portion of an intestine within another portion), twisted intestine, rupture of the intestine, or mechanical stoppage in the bowels from some cause. In all these accidents the pain resulting will be continuous. If the bowels do not act fairly well within 36 hours after an attack of colic, the patient should be given a pint of linseed oil, which may be repeated every 12 hours until the bowels move freely. When recovery occurs it will be noticed that the attacks of pain become less and less frequent until they cease altogether.

## DIARRHEA—SCOURS.

*Causes.*—The causes of this disease are frequently of a constitutional nature, but it may also be caused by feeding too much green and tender grass, new hay or oats, musty hay, in which there is a large amount of the various forms of fungi. Hard driving will often cause it. But whatever the cause, the looseness of the bowels will be due to an abnormal irritation of the intestinal canal.

*Symptoms.*—The animal will show an unnatural looseness of the bowels, which if continued for any length of time will cause it to feel weary, and to lag in the harness. It will often refuse to take its food after a drive; perhaps it will take a chill after being allowed a drink of water. Horses subject to this disease rarely carry much flesh, although they will frequently stand considerable hard work. Symptoms of colicky pains will sometimes be noticed after a hard drive. If driven

long distances in very warm weather, they are liable
to become exhausted and die in a short time. I can
call to mind several such cases.

*Treatment.*—Try to prevent the occurrence of this
condition by changing the diet, and if no improvement
is noticed mix a teaspoonful of bicarbonate of soda with
the food in the morning. This will counteract the tendency to acidity in the intestinal canal. I have cured
many bad cases of chronic diarrhea in horses by this
simple treatment. In some cases however, very little
improvement is noticed. Give a drink of a half a pail
of water, in which has been mixed a handful of starch.
If symptoms of colicky pains are present, give a bottle of
Dr. Heard's Magic Colic Mixture, and allow the horse to
rest a day or two. There are cases in which Dr. Heard's
Condition Powders will effect a cure, but I cannot recommend it as a universal remedy for this disease.

## DISEASES OF THE LIVER.

Serious disease of this organ is very rare in horses, although temporary derangement of function is often present as a result of disease of other organs, or in attacks
of influenza, and from over-feeding with no exercise.

*Symptoms.*—When the liver cells fail to perform their
function, the mucus membranes of the eye-lids and nostrils will show a yellowish tinge, the bowels are constipated, and if long continued, the appetite begins to
fail.

*Treatment.*—Give the animal some exercise if the de-
rangement is due to want of work and over-feeding.
Cut down the allowance of food a half, and give Dr.
Heard's Condition Powders three times a day as directed

on the printed label. This will stimulate the action of the liver-cells, and cause the bile to flow more freely.

## DISEASE OF THE URINARY ORGANS.

HERE again it is rare to find horses affected with disease of those organs. A few years ago I was much interested in the special study of kidney diseases, and I embraced the opportunity of making examinations of the kidneys of all the worn out horses that were killed to be used as food by the animals in the Central Park menagerie. Although I made over 70 post-mortems and a detailed microscopic examination of stained sections of the kidney in each case, I failed to find any with serious kidney disease, the same as is seen so frequently in man. I found one animal that had a cyst that contained over a pint of water, but its physical condition seemed to be fair. The tubules and glomeruli of the kidney in all cases were in fairly good condition. However, we do occasionally have inflammation of the kidneys, as also the condition known as diabetes (sugar in the urine), but although I have made numerous chemical examinations of horse urine, I have never yet been able to find sugar present. I know that the common opinion among horsemen is that horses are frequently the subjects of kidney troubles. I think this is due to the fact that in many cases of disease the urine is only scantily voided ; but this is because absorption from the bowels is not going on, or only very slowly, and the condition of the blood is such that excretion by the kidneys is partly or wholly suspended, and not because the kidneys themselves are diseased.

In this place it will probably be a convenience to the stock-owner if I say a few words on

## POLYUREA—PROFUSE STALING OR PROFUSE URINATION.

—Although not strictly due to disease of the kidneys, the most prominent symptom—that of passing a large quantity of urine—will lead the horse owner to look for a description of this disease under the heading of "Diseases of the Urinary Organs."

*Causes.*—The real cause or causes of the disordered condition of the system in this disease is not well understood. A functional derangement of the nervous system, or a part of it, seems in some cases to be the leading cause of excessive urination. The condition of the blood is frequently abnormal. In other cases we certainly have functional derangement of the digestive system. It is sometimes present as a complication of pneumonia or other lung disease. It is much more common in some localities and in some seasons than in others, which would seem to show that either local climatic or food influences have a great deal to do with its causation. It is said that improperly cured, or musty hay and grain are active causes; we know that several cases will sometimes occur on the same farm.

*Symptoms.*—The one prominent symptom of course is that the animal passes an excessive quantity of urine and has consequently excessive thirst. This is often preceeded by a dull feeling. The animal doesn't drive up as well as usual; there may or may not be a loss of appetite, but it will lose flesh rapidly, and in a day or two will appear to be unusually weak; when in this condition it is peculiarly susceptible to colds, and lung fever is very easily developed. The urine is usually clear—like water.

*Treatment.*—When this condition is discovered, the animal should be immediately taken from work; apply extra clothing to keep the surface of the body warm; the method of feeding should be changed. Put a teaspoonful of bicarbonate of soda in half a pail of drinking water three times a day; give a liberal allowance of water to drink—I usually allow the animal as much as it will take; give it about three quarts of milk at the first morning drink, before the water. Most horses will drink milk when suffering from polyurea. If the milk is refused at the first offer, try again in half an hour, put the pail in the manger if the horse is slow about taking it. Give a liberal allowance of oatmeal water—water in which oatmeal has been soaked for a couple of hours and rubbed between the hands to extract its nutritious elements. Keep the bowels moving by giving warm water and soap injections twice a day. In many cases with this simple treatment the profuse urination will disappear in two or three days. If, however, it still persists, two drams of iodide of potash should be mixed with an equal amount of powdered gentian, made into a ball and given twice daily. This treatment can be very beneficially combined with DR. HEARD'S CONDITION POWDERS. The latter should be administered for some time after the disease has abated, as they are essentially tonic and will greatly assist the assimilation of the food.

## CATARRH OF THE BLADDER AND URETHRA.

*Causes.*—Stone in the bladder may be a cause, but in the majority of cases it is caused by errors of diet, the administration of medicines, exposure to inclement weather, or is a complication of other diseases.

*Symptoms.*—There is a frequent desire to urinate, but at each attempt only a small quantity is passed; frequently, a few drops only, every few minutes. In this respect it differs altogether from the disease last described—Polyurea. There is often considerable uneasiness, which is shown by the frequent changing of the position of the hind feet. The urine is apt to be of a yellow cream color from an admixture of mucus with it. In severe cases the pain will become more severe and continuous, and there will be a loss of appetite, with fast breathing and accelerated pulse.

*Treatment*—The bowels should be kept regular by the use of warm water and soap injections twice a day. The following to be made into a ball and given twice a day: Sulphate of iron, half a dram, calomel, half a dram, gentian, two drams, linseed meal, two drams. Twenty drops of dilute sulphuric acid may be advantageously given in the water three times a day. In cases where the urine is markedly acid, give a teaspoonful of the bicarbonate of soda three times a day in the water. See that the food is clean and sweet smelling. The diet may be changed to advantage in many cases.

It will be often found that this disease is very slow to disappear, but perseverance in the above methods of treatment will in nearly all cases be followed by a permanent cure. If there are calculi or stones in the bladder, it will be necessary to remove them before recovery can take place.

## CONSTITUTIONAL DISEASES.

### RHEUMATISM.

THIS disease is much more common in the domestic

animals than is generally supposed, and is frequently a very stubborn one to treat.

*Causes.*—A sluggish liver is a frequent constitutional cause. There may also be an inherited tendency to the disease. The direct causes are usually connected with such agencies as lower the vital energies, as bad ventilation, errors of diet, hard work, with exposure to great changes of temperature, and atmospheric humidity (moisture). Young animals are more subject to rheumatism than those of mature age.

*Symptoms.*—In mild cases and where local, there will be pain on pressure. If situated in the limbs, there will be lameness, and usually of a rather severe character. We frequently have a swelling of the part affected. The bowels are usually constipated. There is often a tendency in this disease to shift from one limb to another. I have seen two such cases within a month, in one of which it first attacked one limb, then another, and a few days afterward a third. At each new attack the lameness disappeared from the limb previously affected although it was very severe in each limb while it lasted. There was also considerable swelling and intense pain on pressure. There is frequently a great tendency to lie down the greater part of the time. This is very pronounced where we have constitutional rheumatism affecting the whole body. Here we have loss of appetite, high fever, and the early appearance of sores on the hips and elbows of the patient, by constantly lying on the sides. The animal will often refuse to stand long enough to eat a little food. The joints and tendons of the limbs are peculiarly susceptible to rheumatic inflammations.

*Treatment.*—In all cases of Rheumatism, whether local or general, the early administration of the following ball will be beneficial : Barbadoes aloes, 5 drams, ginger, 2 drams. As soon as the cathartic action of this medicine has passed off, give the following ball three times a day : iodide of potash, 1 dram, calomel, 20 grains, gentian, 2 drams, to be continued for several days. Allow as much water as the animal will take. The food should be given rather sparingly, and flax-seed tea made as follows, will be a valuable adjunct : Boil a half a pint of flax-seed in a gallon of water for 15 minutes, and give as much as the animal will take. The administration of DR. HEARD's CONDITION POWDERS will be found very beneficial. Apply a small quantity of HEARD's AMERICAN EMBROCATION to the affected limb, over which should be placed a woolen bandage, this to be continued twice a day until the skin is slightly roughened. Then bandage loosely, without the EMBROCATION for a few days, when, if the animal is still lame, the treatment with the EMBROCATION should be repeated.

## LYMPHANGITIS.— INFLAMMATORY ŒDEMA.— WEED.—MONDAY MORNING DISEASE.

This is a peculiar disease, which has several well marked characteristics. Common bred horses are the most liable to it. It is usually caused by standing a day or two in the stable after an extra hard week's work. This characteristic is so, prominent that it has been named Monday morning disease. It may occur, however, in horses that have had no exercise for several days, and that have had a liberal allowance of food while restin    The hind limbs are much more frequently attacked by the prominent local symptom, than are the fore limbs, and the left hind limb more frequently than the right. In figure 39, we have a good picture of the appearance of a horse suffering with this disease. Although the most prominent symptom is the localized inflammatory swelling, it is undoubtedly a constitutional disease, in which the blood is surcharged with the tissue-

Fig. 39. Appearance of horse with Lymphangitis.

forming element known as lymph, and containing an excess of white blood cells. '

*Causes.*—Over feeding at a time when there is a want of sufficient exercise to use up elements that are carried into the circulatory system. An overflow is nature's remedy for such crowding. The circulation being weakest in the hind limbs, the exudate of white corpuscles and watery parts of the blood takes place in this locality.

*Symptoms.*—The most prominent and diagnostic, is the swelling of one or more of the limbs. This is frequently so sudden that the limb is swollen to double the natural size over night. There is a considerable rise of temperature, the pulse is quickened, and the res-

9

pirations will probably be accelerated. There will be great stiffness of the affected limb and considerable pain when handled. The pain is sometimes so great in the early stages that the animal is covered with perspiration. There is often a marked loss of appetite. The disease has a tendency to recur in an animal that has once been the subject of it.

*Treatment.*—If taken in time —and there is no excuse for delay in the treatment of these cases— a perfect cure may be effected. The following are the guiding principles : allow no food for at least 24 hours, but all the water that the animal will take. The medicinal treatment will consist in giving a cathartic ball made as follows : Barbadoes aloes, six drams, ginger, 1 dram, with injections of warm water and soap until purgation follows. The local treatment will consist in applying hot water to the limb 3 or 4 times a day for about 20 minutes each time. After each bath apply a little of Heard's American Embrocation, which has been diluted with 4 parts of water to one of the Embrocation. As soon as the purging ceases, this should be followed by the administration of the following ball twice a day : Iodide of potassium, 1 dram, calomel, 1 dram, gentian, 3 drams. As soon as the patient can walk around without evincing much pain, it should be given a short walk several times a day. Feed sparingly for about 2 weeks. As soon as the appetite returns and there is no lameness, give gentle work for a short time each day, increasing the amount gradually until a full day's work is well borne. By following this treatment the permanent enlargement of the limb called elephantiasis or milk leg is prevented. We see many cases where the limb is twice the natural size, the enlargement being of a permanent character. When this condition exists the case is incurable. This permanent growth is formed by the organization of the first exudate, in the same manner as granulation tissue is formed. DR. HEARD'S CONDITION POWDERS are an excellent preventive of a return of the malady.

## PURPURA HAEMORRHAGICA, SOMETIMES CALLED BIG HEAD

Somewhat closely related to the last described disease is that of Purpura Haemorrhagica, when we have swellings suddenly appear in different parts of the body. They are very frequent around the head, the lips are sometimes enormously swollen, the walls of the nostrils are sometimes so much enlarged that the animal is threatened with suffocation, and the eyelids are often enormously swollen. The limbs, too, are frequently the parts where large swellings suddenly appear. To determine the exact change in the constitution of the animal exhibiting these local evidences of disease seems to be a question of great difficulty wtth pathologists. It seems to me that writers on this disease have unnecessarily complicated the subject, especially that part of it relating to causes. The swellings are undoubtedly due to an escape of some of the blood elements from the vessels that normally carry the blood. Now, if we remember that the blood is a fluid containing an enormous number of individuals perfectly formed, many of them leading an independent existence in that fluid in the same manner as do some of the animalculae in water, having the power of locomotion, digestion, respiration, and propagation, apparently independent of each other, individually subject to the change known as death, which sometimes takes place in great numbers in a short time without the death of the whole animal: we know that in this disease the coloring matter of the blood (haematin) —which is mostly contained in the red corpuscles—is filtered through the walls of the blood vessels in large quantities, thus showing that there is a disintegration or breaking up of the formed elements of the blood itself. We may therefore consider it purely as a blood disease.

*Causes.*— It usually, though not always, follows some debilitating disease, as distemper and pneumonia. Bad ventilation is often an auxiliary cause, as are also errors in feeding, and very hard work, especially if long continued, where the vital energies are very much weakened.

*Symptoms.*—Sudden appearance of swellings on various parts of the body; the head and limbs rarely escape. There is always some rise of temperature, but usually not to a very high point; the breathing and pulse are nearly always quickened; the animal will show stiffness in the limbs, depending on the extent of the swellings in them. The appetite is sometimes very bad, at other times fairly good. Red spots the size of a copper cent or larger are frequently seen on the inside of the nostrils. It should be borne in mind that the same kind of swellings that occur on the surface of the body may occur internally, in which case the complications may be of a very serious nature. For instance, if the lungs are invaded, pneumonia ensues; if the intestines are affected over any large surface, we may have active purgation. The development of internal swellings is where the great danger lies in this disease.

*Treatment.*—First of all, good ventilation, and plenty of clothing on the surface of the body. For the first day or two allow no food, but instead, mix a pound of oatmeal in a pail of water; after soaking 2 hours, rub the meal well between the hands to impart its nutriment to the water. If the breathing is easy and quiet, give a quarter of a pint of whiskey, mixed with an equal amount of water, 3 times a day. It may be necessary to drench the horse with this. The following ball should also be given 3 times a day: Nitrate of potash, 6 drams, quinine, 2 drams; to be continued until the kidneys are acting freely, when the nitre should be stopped. The following ball should be given once a day, but not with the nitre ball : Sulphate of iron, 1 dram, Gentian 3 drams; to be continued daily for two weeks. After the 2nd day give the horse 4 quarts of milk a day, drenching him if necessary; this to be continued for 2 weeks, unless purging ensues, when it may be necessary to stop for a day or two. If the appetite is fair, a quart of oats with a half a pint of flaxseed may be allowed 3 times a day, after the 2nd day; a little flaxseed tea, 2 or 3 times a day, is a good material for building up the depleted system, and is easily digested. If the bowels are constipated, give injections of warm water and soap 3 times a day.

As to the treatment of the local swellings, there is considerable diversity of opinion whether they should be punctured or not. If the breathing is obstructed by the swellings around the nosrils, they should be punctured in several places, either with a lancet or a common pocket knife to the depth of a half inch. This will allow the fluid to escape and relieve the breathing. Puncturing may also be recommended for the swellings on the limbs and under the chest. It may be repeated every day, for several days in stubborn cases. Hot fomentations are sometimes of very great benefit, followed by the application of Dr. Heard's Dermal Liniment, diluted with 8 parts of water to one part of Liniment. This will prevent a chill to the skin, especially if followed by rubbing with a dry cloth. In many cases the skin will die and slough off in large patches, leaving large raw sores. These should be dressed daily with Heard's Healing Lotion. When the appetite returns, Heard's Condition Powders will be found an excellent tonic for the bloodvessels. Although the successful treatment of this disease often requires great patience, if steadily persevered in, many apparently hopeless cases will be perfectly cured.

## DISEASES OF THE NERVOUS SYSTEM.

### STAGGERS—FITS.

There are several disorders of the nervous system that give rise to the condition known as staggers. True epilepsy may cause a horse to stagger for several paces, when it may suddenly fall and have convulsions for a longer or shorter space of time. We may also have attacks of vertigo or fainting, when the horse will stagger for a few steps, then fall, rising again in a minute or two. This may be due to a variety of causes, as indigestion ; absorption of poisonous substances that are generated in the digestive canal, weakness of the heart, or structural changes in the brain itself. That the most prominent symptoms of this disease, namely, staggering and sometimes falling, with frequently a loss of consciousness, is due to a large number of

different disorders, accounts for the variety of symptoms that are seen in different cases of it.

As an illustration of what may happen in this line
from certain diseases of the heart, I will give a case
that occurred in my own practice.  A horse had suffered
from several attacks of staggers for about six months,
and the night before I was sent for it had staggered
and fell while being driven to the carriage, smashing
the vehicle besides damaging another with which it
had collided.  I began an examination by counting the
pulsations at the submaxillary artery in the usual way.
Waiting 3 or 4 seconds without getting a pulsation, I
was almost afraid the animal was about to have another attack ; but still keeping my finger on the artery,
and noticing the unusual slowness of the pulse, I
took the count for several minutes.  I found that the
number scarcely varied at all from 13 beats to the minute—scarcely $\frac{1}{3}$ the usual number.  I suspected heart
trouble and advised that the horse be kept in the stable
for a few days, saying that I thought it exceedingly
dangerous to drive it in such a condition.  I also said
that it will be likely to drop dead if worked to any extent.  At the end of a week I made another examination.  I advised no treatment.  At the end of six days
I made still another examination.  I found no noticeable
change in the pulse or condition of the horse, which to
all outward appearance was healthy, and the appetite
was good.  I again said that it would be dangerous to
work the animal and gave no hope of improvement.
But much sooner than was expected was my early prog-
nosis fulfilled, for the very next morning, about half
past seven, the coachman came running around to my
house for me, and when we got to the stable the horse

was dead, notwithstanding he had eaten his breakfast as usual that morning. I have never seen a case reported in which there was such a slow pulse, and previous to it I had never taken a horse's pulse where the count was below 28 beats in the minute.

*Symptoms.*—When an attack begins, the first thing usually noticed is that the animal lags, and in a few seconds begins to shake its head. This is soon followed by an unsteady gait. If continued, the animal will fall; sometimes it plunges violently, and cannot be controlled. It may lie on the ground in convulsions for several minutes, and then get up looking very stupid and full of fear.

*Treatment.*—When an animal that is being driven rapidly shakes its head in an unusual manner, it should be allowed to come to a walk. In this way it will frequently recover itself and after that may be driven for several miles without showing any excitable symptoms. A cathartic ball containing six drams of Barbadoes aloes and one dram of ginger should be given about every six weeks to horses subject to this disease. The diet should be kept low, with a quantity of grass when it can be obtained. A teaspoonful of bicarbonate of soda should be given in the food once a day, and to strengthen the digestive system DR. HEARD'S CONDITION POWDERS are the best possible remedy. A horse subject to this disease is a dangerous animal to drive, as he is apt to be suddenly attacked in the most inconvenient places.

## DISEASES OF THE EYE.

THERE are several diseases of the eyes found in our domestic animals, some of which are very rare and will not need to be considered in this little book. There is one disease, however, which is very frequent in horses in this country and will therefore require consideration. I refer to

## CONSTITUTIONAL OPHTHALMIA—MOON

## BLINDNESS—MOON EYE.

*Causes.*—This disease is very strongly inherited. It may appear in the offspring at about the same age that it appeared in the parent, or perhaps earlier. The tendency of this desease to transmission is so well understood in France that the French government studs will not allow one of their stallions to serve a mare that has suffered from it. Bad ventilation in close stables is an active cause. If the drainage is bad the ammonia emanation from the urine will act as a strong irritant to the eyelids. Exposure in stormy weather is also a cause; also pasturing colts on a damp, marshy soil. It is very apt to appear during the teething period. Over-working young animals is a frequent cause.

*Symptoms.*—The eyelids are usually considerably inflamed, swollen and tender; in a day or two there will be a white material deposited over the front of the eyeball, sometimes slight fever and loss of appetite; great desire to rub the eyelids on the side of the stall, and usually a flow of tears over the face.

*Treatment.*—If the patient is run down from overwork, it should have rest and be liberally fed with good, nutritious food. Give internally the following ball: Barbadoes aloes, five drams, ginger, one dram, and remove any irritating cause that may be present. A most important part of the management will be to prevent the animal from rubbing the eyelids against the front or side of the stall. This can be done by turning it around in the stall and hitching a rope to the stall posts on each side, only allowing the horse to stand properly in the stall while feeding. Inject under the eyelids, twice a day—with the syringe that is inclosed with the LOTION—a few drops of DR. HEARD'S EYE LOTION for moonblindness, and continue as long as there is any whiteness remaining on the front of the eye. A horse with this disease should not be worked in the sun or snow, as they are both eye irritants. No time should be lost in beginning the treatment and in nearly all cases the sight will remain good for years, though the animal may be the subject of frequent attacks. To prevent its recurrence, the animal should be kept in the best possible condition,.and nothing conduces more to this end than DR. HEARD'S CONDITION POWDERS, given as directed.

## LAMINITIS—FOUNDER—FEVER IN THE FEET.

A PICTURE of a horse suffering from an acute attack of this disease is seen in (Fig. 40.) In this disease the sensitive parts of the foot, which are in close connection with the folds seen in (Fig. 41) inside the hoof,—are in an inflamed condition. The sensitive parts of the sole or bottom of the foot are also sometimes the seat of inflammation.

*Causes.*—Over feeding is a frequent cause, many cases occurring a day or two after a greedy horse has got loose at night and going to the feed box, has over-eaten itself. A frequent cause is standing in a draft

after driving. This is a common occurrence in summer.
It may follow a chill, which is common enough after a
hard day's work in the late summer, when the coat is
getting long. Evaporation does not take place readily,
and some horses will stand in the stall for hours without the coat drying, even in very warm weather, and
especially is this likely to be the case if there is much
humidity in the atmosphere. Standing long in the
stable without excercise is often a cause, especially of
the chronic variety, as it weakens the circulation of the
foot to stand still for many days at a time. Indigestion
is a frequent cause ; also exposure to very rough
weather.

Fig. 40.  Horse suffering from Laminitis.     Fig. 41. Folds of Laminae.

*Symptoms.*—The common mode of discovery is to go
into the stable in the morning and upon trying to back
the horse out of the stall it is found to be very stiff, and
is made to move with great difficulty. If all four feet
are affected, the animal will be found in the position

seen in (Fig. 40.) If only the fore feet are affected, they
will be put well out to the front, and an attempt made
to stand on the heels. If the hind feet only are affected
they will be carried forward under the abdomen, as seen
in the figure. The breathing will be quickened, the
frequency of the pulse increased, pain in the feet is
often shown by frequently lifting first one foot and then
the other ; especially is this true of the hind feet. The
bowels are usually constipated.

*Treatment.*—Give the following cathartic ball immediately : barbadoes aloes, 6 drams, ginger, 2 drams, followed by an ounce of nitrate of potash once a day,
either in a ball or dissolved in the drinking water,
which should be allowed in liberal quantity. Give injections of warm water and soap 3 times a day until
purging ensues, allow no food for 36 hours, keep the
surface of the body warm with extra clothing.

The local treatment should be as follows : Remove
the shoes from the affected feet, place the fore feet in a
soaking tub filled with warm water and allow them to
remain four or five hours, occasionally removing a pail
of the cooling water and adding hot water, to keep up
a steady heat. Apply poultices of linseed meal, to
which has been added a tablespoonful of HEARD's AMERI-
CAN EMBROCATION. This treatment must be kept up until
improvement follows, and the hot water baths should
be continued till complete recovery. Keep the patient
out of the drafts. After about two days give three
quarts of oats a day for a week, with two or three
pounds of hay, and a few carrots or a small quantity of
grass if in season. Half of an ordinary kerosene barrel
makes a very cheap and convenient soaking tub. When
the acute disease has passed away, a course of DR.

HEARD'S CONDITION POWDERS will improve the general condition of the patient very materially. If properly treated, the average case of laminitis will recover in about a week or ten days. If the disease is extremely acute and hard to get rid of, it sometimes results in a change of structure taking place in the inside of the hoof, whereby the coffin bone is dislocated by being pushed down at the toe, and the sole will be seen bulging downward in a convex position, as seen in Figs. 42 & 43.

Fig. 42. Side view of foot with convex sole.

Fig. 43. Bottom view of foot with convex sole.

If this occurs, complete recovery is impossible. After an attack of laminitis apply a bar shoe seen in Fig. 44 allowing it to set close to the hoof. If the sole has become convex or bulging—sometimes called drop-sole—the shoe seen in Fig. 45 will be the best to prevent

Fig. 44. Bar shoe.

Fig. 45. Shoe for convex sole.

bruising of the diseased sole. There will always be tenderness that will be noticed whenever the horse is trotted out.

# INJURIES.

## WOUNDS.

**HEMORRHAGE.**—In attempting to stop bleeding from a wound, don't use the means ordinarily employed, viz: swabbing with a sponge, as that will only prevent the stopping of the flow. Act as follows : If from a large surface, pack a sponge, or cloths into the wound and retain them there either by hand pressure or by bandage ; and the stronger the pressure the quicker will the bleeding stop. If on slacking the pressure slightly, it is found that the bleeding still continues, apply pressure again, without removing the compress of sponge or rags. In a little while this will stay in place, as it is held by the blood clot. It should not be removed for a little while, so as to give time for clots to be formed in the bloodvessels.

*Treatment.*—In all clean cut wounds where it is possible to bring the edges together, they should be stitched. Various kinds of stitches — sutures—are used by surgeons, depending on the kind of wound they are called on to treat. It will not be necessary for me to go into detail, but to give some plain advice as to sewing up the wounds. It will be necessary first to get the animal under control, and for ordinary wounds the apparatus seen in Figs. 17 and 18 will usually answer the purpose. In fact I sew up many wounds by applying a twitch on the horse's nose only, with one fore foot tied up. I use heavy silk, made specially for sewing wounds ; but when this cannot be conveniently obtained use common strong twine with a small bag needle. The first stitch should be placed so as to bring the edges of the wound together in the middle, going as deep as practical, so that it will not tear out. Now place the second stitch midway between the first and the end of the wound, and so on, always putting the stitch in the middle of the space to be sewn. Bring the edges of the wound close together and tie tightly. It is advisable to leave a small space at the deepest part of the wound for drainage. The

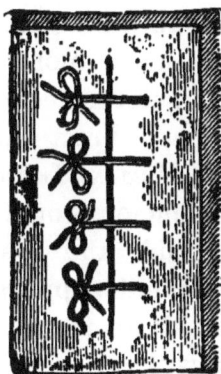

Fig. 46. Sutures for ordinary wounds.

common suture seen in Fig. 46 is the simplest and easiest applied. If there is any dirt in the wound, it must be washed out before being sewn, otherwise no water or dressing should be applied. As soon as the wound is sewn apply on the outside and over the stitches some of DR. HEARD'S HEALING LOTION, as directed on the label. This agent being a thorough antiseptic, will kill all germs that would be likely to cause suppuration in the wound. If the wound is not favorably located for bandaging, the lotion should be applied twice or three times a day and to the parts near it. If it

is situated on a limb, where it is feasible to apply a bandage, the treatment may be varied as follows : after
carefully sewing up the wound and applying the healing lotion as above to the surrounding hair, take a wad
of cotton or oakum and, after saturating it thoroughly
with the lotion, lay it over the wound ; then wind a
bandage over this so that it will cover the wounded
space and retain the cotton pad in position.  If the
edges of the wound have been brought together at all
points, except the bottom part of it and the dressing
applied as above directed it should not be disturbed
for 3 days ; neither should the animal be allowed to lie
down, nor allowed to move more than is absolutely
necessary, as absolute quiet is needed to allow of healing by what is known as "direct union."   Until the discovery of Dr. Heard's Healing Lotion, it was taught by
all authorities that common wounds in the horse except those cn the eyelids and nose, never healed by
"direct union," or without suppuration or the formation of pus.  By the above treatment more than half
the cases will heal without suppuration, and consequent discharge of pus.   On the third day the bandage
should be removed and the lotion applied all around
the wound ; also apply the cotton and bandage as before, after being well soaked with the lotion.   This
dressing may be changed every day until the wound
has thoroughly healed.  The union must become solid
before any considerable motion is allowed, as the young
tissues by which it is brought about will be very tender
and easily torn if stretched.  There is great danger of
the wound opening up again if motion is allowed too
soon.  If there is much pain in the part or great
swelling, with some discharge of pus, the bottom stitch

should be opened and the pus squeezed toward that
part. Apply the healing lotion liberally. Dress twice
a day now. In most cases where the wounds are of
any considerable size, it will be best to keep the horse
tied to a high ring in front of the stall, to prevent motion by lying down or getting up. Most wounds
when treated in the way above described, will heal kind-
ly, but when allowed to run along without treatment,
we frequently have complications which are difficult to
cure, and which in many cases will retard the healing
for some time.

## PROUD FLESH (EXCESS OF GRANULATIONS.)

Whenever the raw surface of the wound projects beyond the level of the skin at the edges, we have an
excess of new material, which will have to be removed
before the wound will properly heal. The healing of
such a raw surface will have to take place by new
growths (additions of new skin around the edges of old
skin) until the new skin finally covers the entire raw
surface. A common cause of proud flesh in a wound is
rubbing or biting it. When such a wound is situated
on the inside of the limb, the animal will frequently
rub it with the opposite limb, causing it to look unusually angry, sore, and swollen, with a reddish discharge from it. In such cases, to cause healing, it is
necessary to prevent mechanical irritation. Irritations
which delay the healing of wounds are very common in
summer, flies causing the animal to rub the part to relieve the itching. To destroy the excessive growth of
repair material, the application of powdered sulphate

10

of zinc is usually sufficiently strong.   This can be repeated every two days or so until the sore is level with
the surrounding skin.   In some cases where the growth
has been going on for some time and is of considerable
size, the most convenient method of destroying it is to
burn it with a red hot iron.   There need be no fear of
damage, provided there is nothing more than the proud
flesh that is destroyed.   It is in all cases necessary to
discover and prevent the irritation that is the cause of
the exuberant growths.   Ordinary wounds that cannot
be sewn up, are best treated by the application of DR.
HEARD'S HEALING LOTION twice a day ; and even in these
cases there is rarely any considerable amount of pus or
swelling in the vicinity of the wound when the lotion is
used.

## WOUNDS OF THE FEET.

These are often of the punctured variety and are usually caused by nails.

*Treatment.*—As soon as discovered the nail or other
sharp implement should be extracted.   In some cases
considerable force will be required.   Having pulled out
the nail, if there is no lameness, nothing more is necessary than the introduction of a few drops of DR. HEARD'S
HEALING LOTION, into the orifice of the wound ; but if
there is much lameness, the horn around the place of
puncture must be pared away by a smith until the bottom is reached by a funnel shaped opening.   This
operation is very important for reasons which will be
easily understood by referring to the article on Tetanus or Lockjaw, (page 97.)   Having made a good large
opening in the horn, drop in plenty of the healing lotion, after which apply a poultice of linseed meal or

some pulpy agent. This should be changed twice a day, and the healing lotion applied liberally before the application of each poultice. Give very little food for a day or two, and if there is any fever, administer a ball as follows ; Barbadoes Aloes, 6 drams, ginger, 1 dram. The poulticing should not be continued for more than 4 days, as it is apt to produce proud flesh in the wound.

## SCRATCHES—CRACKED HEELS.

*Causes.*—These are constitutional and local. What pecular condition of the skin most predisposes an animal to this disease is unknown, when, from any cause, except that the circulation of the blood in the skin on the heel is weakened, it will be much more susceptible to scratches. The most frequent local cause is long continued exposure to very low temperature, as when the animal has to work in ice water during thawing periods, especially after snow storms. This is greatly aggravated by street railroad companies salting the tracks. This not only increases the irritating effect of the street filth, but it lowers its temperature many degrees. The effect of such a material on any part of the body is to lessen the force of the circulation of the blood. If exposed to it for many hours, complete stagnation of blood is apt to occur, and if long continued will cause death of the part. On account of the low vascularity of the connective tissue and structures under the skin of the limbs of horses, this local death from long exposure to low temperature often occurs. The first symptom is swelling near the fetlock, with great lameness, usually followed in a few days by a bursting of the skin and the escape of a foul smelling, pasty material, which is the debris of the dead structures. In other cases a portion of the skin itself is killed by a suspension of the blood circulation, caused by the low temperature. The skin will crack open and the surrounding dead parts ulcerate and fall off as foul smelling debris, leaving an ulcer or ulcers of varying sizes. The destruction is sometimes so great that the tendons and their sheaths become involved, and the results are of a very serious nature.

*Treatment :*—Whenever an animal is found lame and an inflamed swelling appears in the neighborhood of the heels, a poultice of linseed meal, made very soft, should be immediately applied, and changed twice a day. This will have a beneficial effect in various ways ; first, by assisting in restoring the weakened circulation in tissues that are not already dead ; second in causing a softening of tissues that are already dead, and preparing them for removal. In a few days after the attack begins the dead parts of skin will crack and slough off, leaving an open sore. Such wounds often heal very slowly. In order to hasten the reparative process, the poultice, preceded by the application of HEARD'S HEALING LOTION should be continued until the new flesh has grown out even with the edges of the skin, when its use should cease. Now, after applying the lotion every day, a little alum should be sprinkled over the wound. A scab will soon be formed, which should not be disturbed, but a fresh dressing of lotion and alum added every day. A ball of aloes, 6 drams, ginger, 1 dram should be given as soon as the disease is discovered.

## FISTULÆ.

This species of wound occurs in various parts of the body. In most cases on account of the complicated requirements to successfully treat them, I would advise that none but experts should do so.

## ABSCESS—BOILS.

Any local collection of pus in a closed cavity is called an abscess. Abscesses may occur in any part of the body, as a result of local inflammations from blows or other injuries.

*Symptoms:*—A rise of temperature frequently accompanies the formation of abscesses. The enlargements when pressed on, will usually have an elastic feel, and as the pus gets nearer the surface, the swelling will be much softer.

*Treatment:*—When the presence of an abscess is discovered, the sooner the pus is evacuated the better, but be careful that you do not puncture a joint or a tendon. Make a large opening, and as low down as practicable, to get free drainage. This done, the wound should be dressed twice a day by injecting, with a small syringe, HEARD'S HEALING LOTION. The wound must not be allowed to heal outside until the bottom parts heal, so it will be necessary in many cases to insert the finger every day to keep the wound pervious. To assist in bringing an abscess to a head or getting it soft and ripe, a poultice of linseed meal, to which has been added a tablespoonful of DR. HEARD'S AMERICAN EMBROCATION, should be continuously applied.

## WARTS—SKIN TUMORS.

Warts are extra growths of the skin, the causes of which are partly constitutional and partly local. They are of various shapes, some having very broad bases, others having a very narrow base and a full round top, something in the shape of a growing pear. They sometimes grow very fast, especially when the animal rubs them against a hard substance; the outside may even be raw and sore from this cause; in other cases the growth is exceedingly slow.

*Treatment.*—Warts may be removed by the knife or by medicines. Removal by medicines is usually very unsatisfactory, and my advice is not to attempt it. They may also be removed by cutting through the base with scissors, or a red hot iron may be used to sever the base. This will have the advantage of stopping the hemorrhage; in fact, of preventing it. After removal, treat the sore in the manner described for the treatment of wounds.

## ECZEMA.

This is an inflammation of the skin, with swellings that arise quite suddenly, and usually of small size. The inflammation of the skin produced by a blister is a typical eczema. Slight attacks are sometimes caused by the irritation produced by the products of perspiration, especially in the summer months. This will cause the animal to rub against anything that may be convenient for the purpose to relieve the itching. There is also a constitutional tendency to eczema in some horses.

*Treatment.*—Dr. HEARD'S CONDITION POWDERS are of great benefit; also HEARD'S DERMAL LINIMENT, diluting a tablespoonful of LINIMENT with 2 quarts of water; rub on immediately after coming from work, after which rub **dry** with a cloth.

## CAPPED ELBOWS—SHOE BOIL AND CAPPED HOCK

Shoe boil is an enlargement on the elbow, caused by the horse lying with the elbow resting on the heels of the shoe. Capped hock is usually caused by the horse kicking against the side of the stall, or by blows administered in various ways.

*Treatment.*—Apply a small quantity of DR. HEARD's AMERICAN EMBROCATION twice a day. Some means must be devised to prevent the horse from again injuring itself. For shoe boil it may be necessary to keep the animal standing by tying it short to a high ring in front of the stall. As a preventive, the ordinary shoe boil boot is usually successful, if worn every night. If the boil is of long standing and hard, it can only be removed by an operation. When recent, it can be cured as above recommended, if the treatment is persevered in.

### SPRAINS—STRAINS.

A sprain is frequently a serious injury. The muscles, tendons, and ligaments of any part of the body are the most frequent seats of sprains. The most common location in the horse is the tendons on the back of the limbs.

*Symptoms.*—Lameness and tenderness on pressure, followed by swelling and unusual heat in the part.

*Treatment.*—Absolute rest is the first requisite; frequent bathing with hot water, to which has been added a tablespoonful of DR. HEARD's AMERICAN EMBROCATION to each half gallon of water, to be continued for several days in bad cases. Give six drams of aloes in the form of a ball, and feed in limited quantity. If there is still considerable lameness at the end of a week, rub with HEARD's EMBROCATION twice a day; saturate a rag with it and cover the whole with a flannel bandage, to be repeated until the skin is well roughened. If the back sinews are the seat of the sprain, considerable relief will be afforded by the application of a shoe, made thin at the toe, with heel calks about two inches long. This shoe may be worn when first put to exercise after a severe sprain, and as the calks become gradually worn down, the fibres of the tendon will become slowly stretched until they resume their usual length. Allow perfect rest for several days after the animal is able to travel sound.

## SPEEDY-CUT—BRUISING

When the soft structures under the knee on the inside of the leg are struck and bruised by the opposite foot, the injury is called speedy-cut. If the injury be severe, it may cause the formation of an abscess, the pus in which will have to be evacuated before recovery will occur. More frequently, however, an inflammation with swelling ensues, followed in a day or two by the thickening of the tissues at this point. If treated now, the enlargement will be removed without difficulty, but if neglected, and, as often happens, the animal is again ridden or driven, the tendency to bruise by striking is much increased. A new inflammation is set up, and little by little the original swelling is daily added to, until it comes out even with the knee, and has become hard and calloused. This condition is much more difficult to treat, and will require more patience on the part of the owner. Apply DR. HEARD'S AMERICAN EMBROCATION twice a day rubbing well with the hand until the skin is roughened; then cease for a day or two, when the treatment should be renewed. By persevering in this treatment, I have never known a failure to result. Hard, calloused bunches on any part of the body should be treated in the same way.

## WINDGALLS (WIND PUFFS).

Those may be cured by applying a very small quantity of the EMBROCATION and keeping a linen bandage constantly applied while in the stable. This should be continued for a considerable time after the enlargements have disappeared.

## CURB.

This is a sprain of some of the ligaments situated behind the hock, giving rise to a considerable swelling and in some cases great lameness. Its situation is seen at page 56, in Fig. 21.

*Causes.*—Hard riding or driving when young, especially in hilly localities. Inheritance plays a considerable part in the causation of curb in many horses.

*Treatment.*—The animal must be placed in a condition of absolute rest, and DR. HEARD'S AMERICAN EMBROCATION, applied three times a day until the skin is roughened; then allow an intermission for a few days, when if there is still lameness, the treatment should be repeated. This will cure the very worst case of curb. A high heeled shoe as recommended for sprain of the back sinews, will materially assist in curing the lameness.

# THE FOOT.

On account of the great change that has occurred in the foot of the horse during its evolution from a five-toed

to a one-toed animal, there is constant deviation from
any one absolute, special, and exact form and shape.
The foot of the horse is in a condition similar to any
structure that has been developed in comparatively recent times—where change readily takes place—and
where the normal balance is readily upset. A glance
at the following diagram will show the changes in the
foot of the horse. Fig. 47.

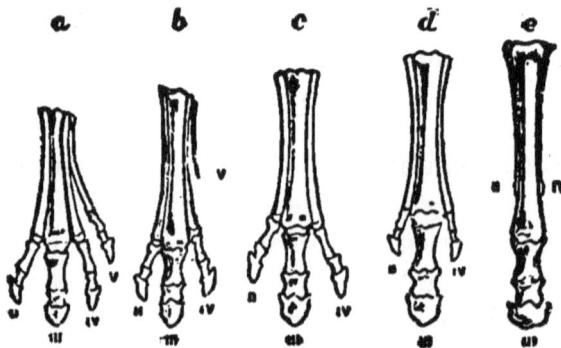

Fig. 47.  Showing feet of ancestors of Horse.

Again, the uses to which horses have been put by
civilized man have necessitated changes of structure
in many parts of the body, notably in that of the foot.
For instance, travel on macadamized roads requires
a foot structurally much stronger than on soft ground.
In other words, the foot that will suffer no disorder
when used for travel on soft, dirt roads, may be irrem-
ediably damaged by travel for only a short time on the
macadamized roads of to-day. It has been frequently
argued by writers on the foot that horses in a state of
nature rarely have foot disorders, and that therefore
shoeing is the cause of most of the ills that affect the

foot. But we should remember that the conditions under which the horse of to-day has to labor are totally different from those in which the horse is found in a state of nature. It has therefore been found necessary to use various artificial means to preserve portions of the animal body from partial or total destruction. In very early times it was found necessary to protect the foot of the horse, and up to the present time a rim of iron has been deemed the most serviceable protection for the bottom of the foot. But this takes no account of the outside of the hoof which is also subject to a considerable change on account of the following departure from natural conditions. In a state of nature the horse traveled on pastures continually. These pastures were kept moist by frequent rain and dew, thus keeping up a water or moist saturation the greater part of the time, thus enabling the horn to grow tough and strong and preventing any undue dryness, except in unusually dry seasons, or in unusually dry and arid places. Now we have the horse traveling much of the time on very dry roads, covered with dust which assists in absorbing the little moisture that may be near the outside of the hoof, and at night a very large number of horses are kept in stables where everything is dry under the feet—just the opposite of natural conditions. Evaporation goes on continually from the surface of the wall of the foot. The horn becoms dry and brittle, is subject to cracks and bruises, in fact, almost the reverse of the tough, moist horn found in the hoof of the horse kept under natural conditions. To counteract this tendency to dryness of the horn, various applications have been tried, and all with varying success, until by experimenting with various agents that had been recommended by previous writers, I came to the conclusion that a mixture of Nonane, Hexadecane and certain fixed oils, when applied daily to the horn, made a perfect protection from evaporation of moisture and consequent brittleness of the horn. It is now made and sold in large quantities as DR. HEARD'S HOOF LINIMENT, and should be used by everyone who has a horse that is of necessity kept from natural pastures, where the foot would be kept moist by rain and dew. The advantage of using this liniment instead of the ordinary tar compounds, will be readily understood when it is considered that tar is an acid substance, which absorbs and holds sand and dirt in large quantities when brought in contact with it.

## CONTRACTED HEELS.

This is a condition in which the hoof is smaller than normal, and may exist with or without lameness. The effect, however, is to cause a pressure on the bloodvessels and nerves contained in the sensitive structures inside the horn.

*Causes.*—Inheritance is a frequent cause. Huidekoper says that a horse that is contracted, though not lame, should be excluded from the stud. Hot dry weather tends to diminish the quantity of moisture in the horn and predisposes to a shriveling of the horny fibres. We have already referred to the bad effect on horn of keeping horses in the stable. This is aggravated if the animal is kept in the stable without exercise for any great length of time. The circulation of blood through the foot will be diminished, as will also the amount of horn moisture. The practice of many smiths in filing away the gelatine coating that forms on the outside of a healthy hoof, is also to be deprecated, as it removes the natural protection against excessive evaporation. Opening up of the heels is also a great cause of curling in of the heels, for if the horn was not removed, it would tend to act as a wedge to prevent the curling inward.

*Symptoms.*—A pointing of the contracted foot is usually an early symptom, as is also a gradual curling in of the heels of the hoof toward the frog. If lame, the lameness is usually most severe when first coming out of the stable.

*Treatment.*—Should be both preventive and curative. To prevent contraction, colts and horses should not be kept in stables for days at a time where the floors are of dry boards. The hoof requires exercise and soft, but clean boards. This has especial reference to young farm horses that do not wear shoes. The application of DR. HEARD's HOOF LINIMENT is an excellent preventive of the dryness and brittleness that leads to contraction. Proper shoeing is also a preventive. The kind of shoe which most frequently causes contraction is that in which the outer rim of the heel is higher than the inside. The heels in this case will be pressed inward toward the frog at each step the horse takes. I have seen many shoes made and fitted with this evident defect. The bearing of the shoe at the heel should be perfectly level, if it is meant to be applied to a healthy foot.

*Curative Measures.*—Various devices have been invented to spread contracted heels, one of the most simple being an instrument which works on the principal of the jack screw. As to the efficiency of this instrument as an expander there is no doubt, but its use is founded on the error that the foot of the horse is a non vital piece of machinery.

Some years ago one of my employers owned a horse that was the subject of contraction, and having heard high recommendations of the use of this screw by a professional operator, he asked my opinion of it. I was compelled to say that unless such force was used as would cause a temporary laminitis, it would take longer to expand the contracted heels in this manner than by other and less vigorous measures. The ease with which the heels could be forced open by means of this jack screw appeared so plain that the gentleman concluded to give it a trial; and I had the opportunity of watching the results, in which I confess I had some curiosity. I found that immediately after the screw had been used, the horse appeared to have relief for a few minutes, but that it was soon followed by increasing pain to such an extent that one day when I happened in the stable (about an hour after the operator had gone), I exclaimed, "Why your horse is foundered!" The attendant informed me that the horse always stiffened up in that way about an hour after being operated on. I was not surprised, as I had told the owner (when my opinion was asked) that, "on account of the close vascular connection between the heels of the coffin bone, which could not be made to expand, and the heels of the hoof—which would be pushed apart by the use of the screw—some of the bloodvessels would necessarily be ruptured, and the remainder would be stretched so as to be considerably weakened, and thus allow blood to be extravasated and an inflammation set up." This was precisely what had happened in the above mentioned case, and sure enough the horse was foundered, but a part of the treatment being the immersion of the feet in a tub of hot water for 4 hours each day, the acute symptoms were relieved by the time of the next operation—two days after. I mention this case because at first sight, and without a knowledge of the minute anatomy of the foot, it would appear that the use of the jack screw would accomplish expansion of the heels much more completely and easier than any other means. I have no doubt that there are at present men traveling through this and other countries making money by the use of such injurious devices, for the field for their labors is unlimited.

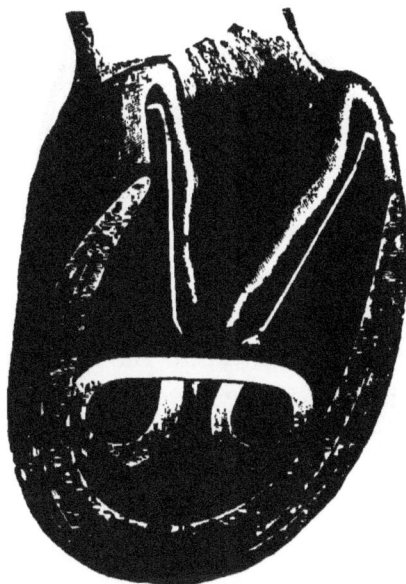

Fig. 48. Dr. Roberges' Spring inserted in Hoof.

Another ingenious device has been invented by Dr. Roberge, Fig 48. It consists of a steel spring, which, when inserted inside the heels, exerts a steady pressure on them in an outward direction. As this pressure is not severe and is continuous, no bad effects follow the use of the spring, but it is continuously exerting an expanding pressure on the parts to be spread. This invention can therefore be highly recommended.

After all, expansion of contracted heels can be accomplished by shoeing, and in a very simple manner. The kinds of shoe are shown in Figs. 49 & 50. It will be seen

Figs. 49 and 50.  Expanding shoes for contracted feet.

that they are plain shoes, differing from the ordinary shoe by having a slant or bevel at the heels in an outward direction, instead of being precisely level. The extent of this slant or bevel is regulated entirely by the smith who makes the shoe, and may be either slight or of considerable extent. If the contraction is slight, the bevel may be shallow, whereas if the contraction is very great, the bevel may be at a more acute angle. The care necessary in the application of this shoe is to be sure that the inside edge of the heel of the shoe shall be set inside the contracted heel, so that every time the horse steps the heel will be pressed outward by the bevel of the shoe. This shoe should not be fitted exactly tight to the heels, as they drop slightly each time the animal places the weight of the body on the foot. It is also necessary to remove the shoe about every two weeks to have it refitted, as the heels will be expanding while the quarters of the foot are held in one position by the nails.

## CORNS.

Corns in horses feet are of an entirely different nature from those of the human feet. In horses there is an escape of some portions of the blood from the blood-vessels located in the sensitive sole of the foot. There is, in fact, a local congestion or inflammation of the part affected.

Fig. 51. Contracted foot showing seat of corn.

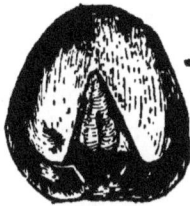

Fig. 52. Flat foot showing seat of corn.

Fig. 53. Testing foot for tender sole.

*Causes.*—Contracted heels, wearing shoes for too long a time without removal and concussion on hard roads are all prolific causes of the bruising which produces the escape of blood from the vessels. To test for tenderness in any part of the sensitive sole use a smith's

11

tongs as shown in Fig. 53. A deep-seated and fresh corn will often be discovered by this means when no superficial discoloration can be seen.

*Treatment.*—If the heels are contracted, they should be expanded in the manner recommended for the cure of contraction. If the shoes have been allowed to remain on too long, they should be removed more frequently. To prevent bruising the shoe should not be fitted too tightly to the heels. A small open space should be allowed between the heel of the shoe and the heel of the foot. The application of DR. HEARD'S HOOF LINIMENT will materially assist the horn to become tough and elastic, when the concussion or jar to the foot will be greatly lessened. Unless the lameness is severe, gentle exercise is better than standing in the stable. If the lameness is severe, showing that the local inflammation is also severe, it is best treated by poulticing with linseed meal, to which has been added about a tablespoonful of DR. HEARD'S AMERICAN EMBROCATION. If the animal is turned out to pasture, the shoe called a tip, Fig. 54 should first be put on to prevent the wall of the hoof from being broken away.

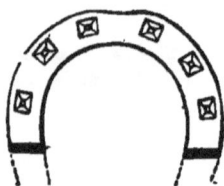

Fig. 54. Tip.                    Fig. 55.  Hoof with Quarter Crack.

## QUARTER CRACK.

This is a separation of the fibres of the hoof from

their connections with each other, occurring at the quarters.

*Treatment.*—The application of a bar shoe—round shoe, Fig. 44, and making a grove into the horn across the direction of the fibres at the top of the crack with a thin hot iron, as seen in Fig. 55, to prevent the further extension of the splitting as the new horn grows down. To stimulate the new growth of horn, apply DR. HEARD'S AMERICAN EMBROCATION, daily to the coronet over the quarter where the crack is situated, until the skin is roughened. After an intermission of a few days, repeat the treatment. Use DR. HEARD'S HOOF LINIMENT to prevent brittleness and a dry condition of the horn.

# TEETH.

## CUTTING TEETH.

When young horses are cutting their teeth, they are frequently the subjects of great nervous tension, and are predisposed to become easy victims to numerous ailments. Many cases are met with in which the second teeth are cut before the first teeth (which they will replace) are shed. In these cases the first teeth ought to be removed so as to allow the second teeth to come up in their proper place, otherwise they are apt to be crowded out from the normal position, and leave the mouth somewhat deformed. It is a very easy matter to remove the first or milk teeth at the time of the cutting of the second permanent teeth, as the root has been mostly absorbed, and the tooth can be extracted by any kind of pliers, when no proper tooth forceps are handy.

## WOLF TEETH.

Another frequent annoyance among horses—and colts are as frequently the subjects of it as old horses—is the presence of wolf teeth—also known as Remnant teeth.

They are found immediately in front of the first upper molar, and they are the only remains of what was in the early ancestors of the horse the first molar. To discover their presence it is only necessary to hold the tongue with one hand and run the thumb of the other hand back against the first upper molar, and if a wolf tooth be present, a small tip probably no larger than a pea, will be felt right in front of the root of the first back tooth. They will be occasionally found an inch or two in front of the first molar, but of course behind the tush or bridle tooth. In some cases they may be felt beneath the gum before they have cut through.

*Symptoms.*—Wolf teeth often cause a horse to drive badly, especially when an upper-jaw bit and check rein is used, as the bit will be brought right back in contact with them, and apparently cause the animal much pain. They often cause a horse to slobber a great deal when being driven.

*Removal.*—Various forms of forceps have been invented for extracting these teeth, and when easily obtained should be used; but as stock owners as a rule cannot conveniently command the use of dental instruments, nor the services of a skilled dentist, a simple means of removal may be mentioned, namely a blunt chisel. Place it against the base of the tooth and strike a slight blow with a hammer or mallet. The tooth will usually become loosened, and if it does not drop out, will be easily removed by the finger and thumb. The chisel should be so placed that it will glide by the first molar when the wolf tooth is struck. It is rather a crude method of removal, but is often the most convenient obtainable and quite as effectual as any other.

## MOLARS.

Another anomaly is also frequently found in our present domesticated horse, viz., the outer edges of the upper molars are found to overhang the edges of the lower molars, as seen in Fig. 56. The cause of this seems to be that with our present method of feeding, where the horse often gets through a meal in a half an hour or less time, there is not sufficient wear of the teeth to keep them level. A question frequently put to

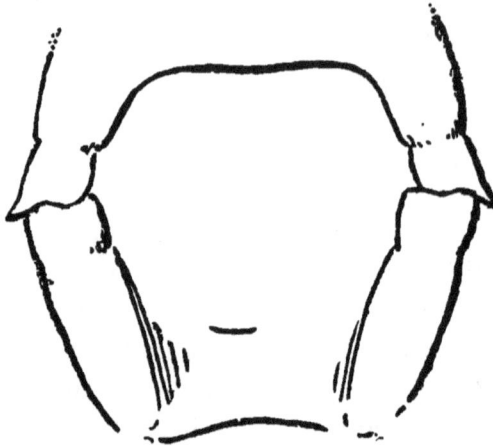

Fig. 56. Teeth showing overhanging upper molars.

me by employers is "What does the wild horse in a state of nature do about getting his teeth filed?" The answer to the question is that, as the wild horse is about all the time grinding food that he has to search for so diligently, the teeth are worn sufficiently fast to keep them level. By looking at the above picture we can easily see that the effect of these overhanging edges is to cut the cheek while feeding, or more frequently perhaps when the bit is placed in the mouth and held by the reins; the cheek is now puckered up and will be brought with some force against the sharp edges of the teeth. It is no uncommon thing to find horses with large sores on the inside of the cheek that have been caused by these sharp edges of the teeth. To examine a horse's mouth for these sharp edges it is only necessary to hold the tongue with one hand and run the thumb of the other hand back along the outside of the upper teeth, when, if any sharp edges or points are present, they will be immediately felt. Having discovered the sharp edges, it will be necessary to level them down with some kind of rasp. Different kinds are used, but one of the simplest and most convenient will be found pictured in Fig. 57. This should be rubbed on the edges of the teeth until they have a level feel when examined by the thumb as above described. Care should be taken that the last molars are reached with the file, and that they are made level as well as those placed more toward the front of the mouth. Nearly every horse that is fed in the ordinary way on oats and hay will require to have his teeth dressed about once a year.

Fig. 57. Tooth rasp.

## *THE GROWTH AND WEAR OF TEETH.

Healthy horses' teeth (the second set, not the first), grow practically throughout life, but much slower after fifteen or sixteen years of age than before. This growth is designed by nature to counteract the enormous wear of the teeth, the horse having to perform for himself that which the miller performs for man. If the lower molar, illustrated in figure 59, had met its corresponding upper molar, its two inches of extra growth should have been worn off by attrition (mastication) and a like amount from the upper tooth. But the upper tooth was unfortunately lost. The lower tooth therefore, grew till the friction from it on the upper jaw killed the horse. Various kinds of instruments are used to remove this extra growth of teeth. A moderately sharp chisel may be used, but there is danger of its slipping and cutting the soft structures of the mouth or throat, for it has to be struck with considerable force. When convenient send for a veterinary den-

Fig. 59. Back lower molar; extra growth begins at dotted line.

* This page is from "Horse's Teeth," by W. H. Clarke.

tist. There are very few animals besides the horse 'whose teeth grow throughout life.

There are many diseases to which the teeth are subject, but we shall only notice one, viz.:

## ULCERATION OF TEETH—NECROSIS—ROTTEN TOOTH.

Fortunately the horse is not as liable to necrosis of the teeth as is man ; in fact, it is rather rare except in very old animals. In many cases horses undoubtedly suffer from ulceration of the teeth and the consequent pain for some time before discovery. The animal will lose flesh, the hair looks dry and there is a general appearance of poor nutrition. In advanced cases there is sometimes swelling of the glands under the jaw, with foul-smelling breath ; a little later, if an upper tooth is the subject of ulceration, we are likely to have a foul-smelling discharge from the nostril of the affected side. I have known this condition to continue for many months without the true cause being discovered.

The mouth should be examined by inserting the mouth speculum, shown in figure 60. The tongue

Fig. 60. Mouth speculum.

should be held by an assistant and the head kept steady
by another assistant, holding one ear with one hand,
and the nose with the other. The operator can now
run the hand back through the opening in the speculum, and examine the teeth carefully. If ulceration has
progressed to any extent, he will feel the hollow space
(cavity) in the tooth, or perhaps between two teeth.
Having discovered this hollow space, the diagnosis is
certain. I shall not describe the methods of removal
of the remaining portion of the ulcerated tooth, as the
services of a skilled operator with complicated instruments will in nearly every case be found necessary. If
an old horse is the subject of the disease, it is best that
he be destroyed.

# PARASITES.

Parasites are living organisms that obtain their substance from the nutritious material contained in the
bodies of other living organisms. Some of the parasites are of a vegetable nature, as the fungus of the
ringworm, occurring on the skin of man and most of
the domestic animals. Many parasites belong to the
animal kingdom, as worms, ticks, lice, fleas and various
kinds of flies. Then again some of the most destructive parasites that infect man and animals are neither
vegetable nor animal, but seem to have a realm by
themselves, about midway between the vegetable and
animal kingdoms. To this class belong the various
species of bacteria and cocci that are the cause of
most of the infectious and contageous diseases in man
and animals.

We shall not attempt to describe this last class. For

their effects we must refer the reader to the various infectious and contagious diseases mentioned in another part of this work. We shall only consider the most common vegetable and animal species and their effects on the system.

## RINGWORM.

This is a disease of the skin and is caused by a peculiar plant, seen in Fig. 61. This plant causes ringworm in horses. Fig. 62 represents a plant which causes ringworm in poultry. Most of the domestic animals

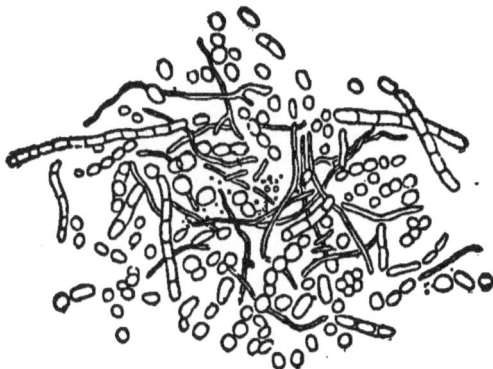

Fig. 61. Parasite causing ringworm in horse.

are subject to the invasions of this parasite, and I have noticed that in damp climates or in unusually wet seasons, especially in hot weather, the disease is much more common than in cold or dry climates or seasons. There are several species of vegetable fungi which cause skin diseases, each giving symptoms somewhat different from the other, but as they will require about the same kind of treatment to destroy the plant, I shall describe them all as ringworms.

The most common form is where the mode of growth
of the fungi is in the form of circles of greater or less
regularity. These fungi attach themselves to the skin
and burrow down into the sheaths of the hairs, causing
more or less irritation and inflammation. They live and
develop by absorbing the nutritive elements of the
skin. Young animals or those poorly nourished are
more frequently the subjects of the disease than older
and stronger animals. In calves it is found most frequently on the eyelids and lips and skin of the neck ;
rarely on the hind limbs. In horses it is most frequently found in the region of the root of the tail, and seems
to be caused in many cases by the crupper of the harness carrying the spores or germs of the disease. It is
quite common in dogs, and may occur in any part of
the body, causing an intolerable itching.

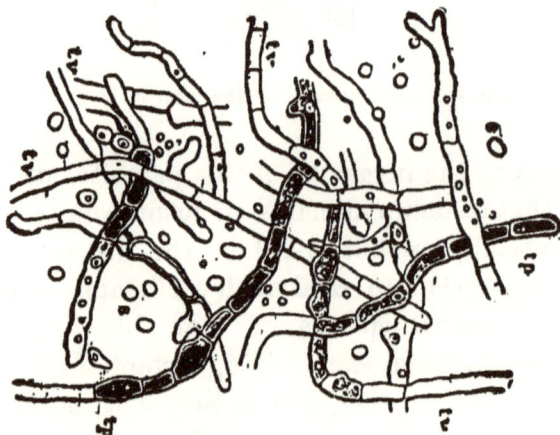

Fig. 62. Microscopic appearance of parasite causing ringworm in poultry.

Perhaps the worst feature about this disease is that children—and sometimes adults—frequently take it from the horse, calf or dog.

Grooming utensils, clothing, harness, etc., often convey the parasites from one animal to another. I believe that at least 10 per cent. of the horses that are brought here from the west, become affected with ringworm before they have been here a year.

*Treatment.*—The treatment for ringworm in horses is usually not a very difficult matter, yet there are some cases that are unusually obstinate. In most cases a few applications once a day of Dr. Heard's Mange Cure will be sufficient to completely destroy the parasite. It should be used full strength, by rubbing a little of the salve into the spot with the fingers. In a few days the young hairs will be seen sprouting out of the skin. This is the proof of cure. If the parasite has burrowed down into the hair sheaths and sebaceous follicles, it will be more difficult to get at the spores, and the disease will persist longer and will require a longer treatment, sometimes for 2 or 3 weeks. But the important point is to persist in the treatment without intermission until the parasite is destroyed which it will certainly be, no matter how severe the case, if the daily application of the mange cure is steadily continued. The treatment for calves is the same as for horses.

For dogs the mange cure should be applied twice a day by dissolving a tablespoonful in a pint of warm water and rubbing in with a brush. After the first week, once a day will be sufficient. It should be remembered, however, that in dogs the disease is very persistent, on account of the spores being deeply imbedded in the skin. The treatment will sometimes have to be continued for a month or six weeks, although I have often cured recent cases in 2 or 3 applications  I have never yet found a case that has resisted this treatment when I have personally supervised the applications.

The utensils and stalls should be thoroughly disinfected by being boiled if possible.  The immovable parts of the stable fixtures should be well washed with hot water and soap and then coated with a solution of the MANGE CURE—a half pint to two gallons of warm water.

## FLIES AND MOSQUITOES.

In some climates these parasites are exceedingly deleterious to the health of domestic animals.  In Fig. 63 we give a picture of the business end of a mosquito, in which it will be seen that it is built especially for the purpose of boring into tough structures.  The sharp pointed dagger is inserted through the skin of many animals with the greatest ease.

Fig. 63.  Head of Mosquito showing the sharp proboscis.

There is a fly known as the grey horse-fly, in some localities as the green-head which is very annoying to horses with sensitive skins. The common fly too is very annoying to many animals in the latter part of the summer. The annoyance and irritation caused by the parasites tend to keep many horses poor in flesh, reduce the amount of milk yeilded by cows, and when they are exceedingly numerous, often act as predisposing causes of other and more serious disorders. We shall not in this place notice the gad, bot, or grub-fly, but describe each under the heading of the disease it causes,

*Treatment.*—Keep stable and other buildings where animals are housed as dark as possible; use screens to prevent the entrance of flies when the windows and doors are open. When at work where these pests are especially annoying, fly nets may be advantageously used. Twigs or the small branches of trees attached to parts of the harness are very useful as preventives. The application of a solution of walnut tree leaves once a week is said to be an excellent preventive. The leaves should be boiled for a few minutes; but the animals may be simply rubbed with the leaves themselves.

## FLEAS.

This order of parasite is very annoying to many animals. Drawings of the heads of some of them are shown in Figs. 64, 65, 66 and 67.

Fig. 64. Head of man flea.       Fig. 65. Head of dog flea.

Fig. 66.  Head of rabbit flea.　　　　Fig. 67.  Head of fowl flea.

The above illustrations show the heads magnified 30 diameters.

*Treatment.*—Dr. Heard's Mange Cure, applied as directed.   Persian Insect Powder will drive the fleas off, but will not kill them.   Kennels should be frequently treated with boiling water.   The floors should be sprinkled with Creoline Powder, which will also drive the fleas away.

## CHIGOE—JIGGER—SAND FLEA.

Fig. 68.   Chigoe or sand flea.

In Fig. 68 we have a representation of this insect, which is such a pest to man and animals in Texas, Kansas, and many Southern States.   It attacks sheep, goats, horses, mules, cattle, cats, dogs, and pigs and chickens are its especial prey.   The female of this insect penetrates the skin of man or animal, remaining imbedded for from three to seven days. It contains from 150 to 250 eggs. These are implanted and sometimes hatched in the pustule that is formed by the irritation which the parasite sets up.

*Treatment.*—To prevent the access of this insect, the body should be sponged with a solution of a tablespoonful of DR. HEARD'S MANGE CURE in a gallon of water about twice a week. When the parasite is already in the skin and the pustule formed, it should be punctured with a needle and dressed twice a day with Dr. HEARD'S HEALING LOTION. This will prevent the developement of the eggs which are nearly always present and may cause dangerous symptoms if not destroyed.

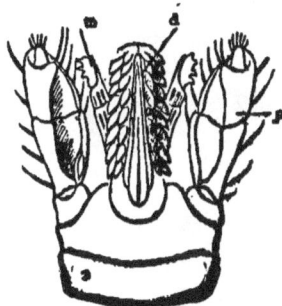

Fig. 69. Head of tick enlarged 50 times.

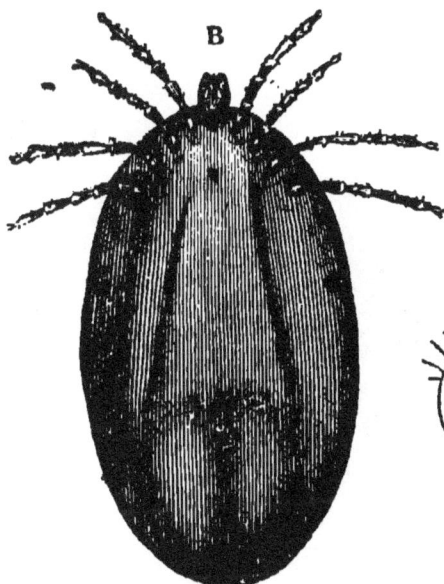

Fig. 70. Dog tick enlarged after feeding

Fig. 71. Dog tick natural size

12

In Fig. 69 is seen the head of one of the ticks, enlarged 50 times, and in Fig. 71 is seen a dog tick of natural size and in Fig. 70 same enlarged after feeding. These parasites do not seem to cause a great amount of suffering as nearly all sheep seem to be affected with them without ever noticing their presence. Each species of domestic animals has its peculiar species of tick, each being different from the other.

## LICE.

Most of these parasites have the mouth arranged so as to act as suckers, and many of the species have only one claw on each foot, as will be seen on referring to Fig. 73.

Fig. 72. Horse louse.

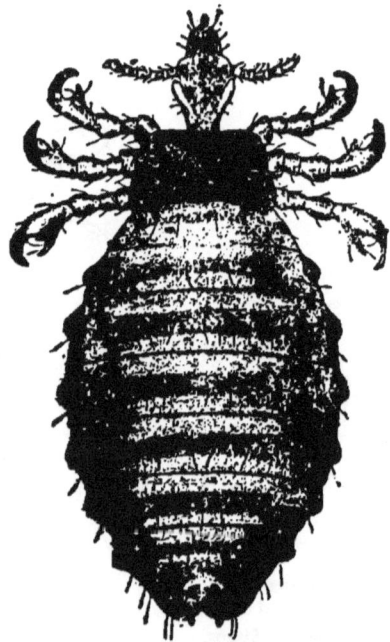

Fig. 73. Ox louse.

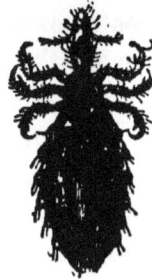

Fig. 74. Sheep louse.      Fig. 75. Goat louse.      Fig. 76. Pig louse.

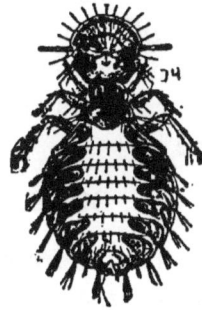

Fig. 77. Dog louse.      Fig. 78. Fowl louse.

They are propagated by eggs. These are known as nits, and in many species of animal are found attached to the hair by a gluey substance.

*Symptoms.*—All animals reveal the presence of these parasites by rubbing and scratching. If long continued, the skin will be bare in spots, and a little later sores will be produced. Such animals are usually in a bad condition, unable to stand a hard days work. Long hair, lack of cleanliness, and general debility all conduce to the multiplication of lice.

*Treatment.*—Cleanliness is a paramount necessity.

In short-haired animals it is comparatively easy to get rid of lice. In animals with long hair a cure will be greatly expedited if the hair is removed. This will also prevent the nits from having a suitable place for deposition. The use of DR. HEARD's MANGE CURE of the strength of a tablespoonful to a pint of warm water, will kill the lice, but it will be necessary to repeat the treatment about every 4 days to destroy the parasites that have been hatched since the last application. When a large number of animals are the subjects of a plague of lice, it will be advisable to disinfect the stables or buildings by washing either with 5 per cent creoline solution or with pure vinegar once a week for 3 or 4 weeks. A decoction of 3 ounces of tobacco, boiled in a quart of water, is also a good remedy when applied every 4 days.

## SCAB, ITCH, MANGE.

These parasites are a very serious scourge to some of the domestic animals, especially to sheep. Man is also occasionally the subject of this pest, in whom it is quite difficult to cure. Drawings of the insects which cause mange in the horse, scab in sheep and itch in man are seen in Figs. 79, 80 and 81.

It will be seen that they are all supplied with a boring apparatus at the point of the head, and also suckers at the ends of the extremities or legs. These suckers are especially brought into use when the female enters the skin, for only the female enters it. They hold the insect fast to the skin, and the point of the nose is pushed through in much the same way that a mole enters a burrow in the ground. Once inside the outer skin, the burrowing continues. The eggs are gradually deposited along the burrow until about 15 have been left in the gallery at different intervals, as is seen in Fig. 82 In from 36 to 100 hours the young insects are hatched and begin their work of destruction. There are said to be twice as many females hatched as males, which is in accordance with the well known law that where the

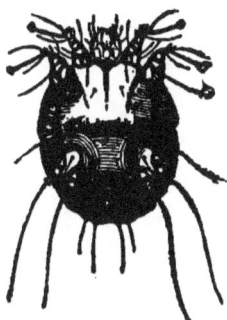  

Fig. 79. Mange insect.  Fig. 80. Scab insect.  Fig. 81. Itch insect.
Horse.                   Sheep.                  Man.

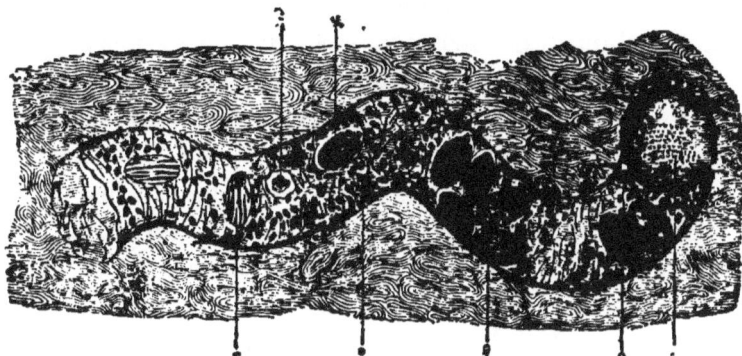

Fig. 82.  Gallery or burrow under the skin made by the female mange parasite (showing eggs along the track of the animal.)

conditions of animal life are favorable, the females always outnumber the males. In this case nutrition is in abundant supply. The eggs will retain their vital properties from 1 day to 3 weeks, depending on the temperature after removal from the body. Hence the great liability of a recurrence from reinfection after the parasites themselves have been destroyed.

*Symptoms.*—In all animals the invasion of this insect causes intense irritation and prevents ordinary and necessary rest. They lose condition and from frequent rubbing will soon be covered with sores as the disease will extend in various directions. In sheep, hard crusts or scabs are soon formed over the spot where the parasite is located. The loss from this disease is probably not very serious in this country, except in the case of sheep in which it is enormous. Many parts of this country that are well adapted to raising sheep are almost entirely devoid of this very profitable animal because of the difficulty of preventing infection. Many farmers in Kansas have informed me that the great losses from scab in sheep have compelled them to abandon an otherwise profitable industry.

*Treatment.*—In mange of the horse the affected animal should be isolated. No grooming instrument used on the infected animal should be used on the uninfected ; neither should the same harness be used. The affected horse should be clipped and kept at some distancé from the stable and the hair burnt. The horse should now be well covered with soft soap and rubbed with the hand to soften the scabs. At the end of 2 hours the animal should receive a thorough scrubbing with a brush and warm water, after which it should be dried with whisks of straw or cloth, the latter being burnt, as they will contain both the parasites and their eggs. Now a 15 per cent. solution of creoline in water should be applied, and the treatment repeated 5 days after, two or three times successively ; or the following application may be used instead : 2 ounces of tobacco, boiled in a quart of water for 15 minutes, and repeated every 4 or 5 days. As this solution does not destroy the vitality of the eggs, and as they hatch out in a few days, repetition of the application will destroy each new brood.

*Treatment for Scab in Sheep.*—When only a few sheep are affected, and the invasion is recent, the application 3 or 4 times at intervals of 4 or 5 days of the tobacco solution mentioned above will be sufficient to effect a cure ; but if, as frequently happens, a whole flock becomes infected, the successful treatment will be a problem of much greater difficulty, on account of the impossibility of isolating those that may be easily cured from those in which the disease is deep-seated and hard to cure. The following will eradicate the scab from a large flock when fully carried out. The most economical treatment for a large number of animals will be by using a dipping fluid, to make which there are numerous receipts. The cause of the non-success of most of them is that the details of application are rarely carried out. If the wool is long it will be much more difficult to effect a cure than just after its removal. The sheep to be treated should therefore be shorn before the treatment is begun. The French use a solution of creoline in water in the proportion of 1 pint creoline to 10 gallons of water and a pint of glycerine. In Great Britain the preparations of arsenic are mostly used, a formula of which I will here give. Arsenic 35½ ounces, sulphate of zinc (commercial) 11 pounds 1½ ounces, aloes 1 lb. 2 ounces, water, 22 gallons.

Dissolve the arsenic in 4 gallons of boiling water ; also dissolve the aloes and sulphate of zinc in 2 gals. of cold water, then gradually mix the 2 solutions, and add the remainder of the water. This will not be an expensive dipping solution, and is considered sufficient for dressing 100 sheep. It does not stain the wool. The dipping fluid should be kept continuously warm while the dipping is going on. A large tub is required for the dipping process, and if a large number are to be dipped, four men should be employed, one to get the sheep as they are required by the men at the tub. Three men should be at the bath, two of whom must be supplied with brushes to scrub the sheep with while immersed in the bath. As fast as the sheep are completely dipped, they should be passed to another inclosure, to prevent them from mixing with the undipped.

No man with sores on his hands should handle arsenical preparations. The sheep should be kept immersed in the bath for about 2 minutes, care being taken to keep the nose and mouth out of the solution. A solution of tobacco in the same proportion as recommended for mange in the horse is a cheap and effective dip, but it should be repeated 3 or 4 times at intervals of about a week. The arsenical bath should be repeated in about 2 weeks. The sheep should not be allowed to return to the same pasture for 6 or 8 weeks if the temperature is at all high.

## WORMS.

The bots and stomach grubs are the most frequent

in the horse and are well illustrated in Fig. 83. The origin of the bot is traced to the gad fly, seen in Fig. 84.

Fig. 83. Bots.

Fig. 84. Gad fly.

During the later summer months these flies may be seen about the knees and legs of horses at pasture, laying their eggs, which are carried into the stomach by the horse licking the hair. On arriving in the stomach, which is the natural habitat of the grub, they fasten themselves to the mucus membrane, and there remain for perhaps several months. In some cases they are so numerous as to materially interfere with the process of digestion, and the horse will lose flesh and suffer undue irritation of the stomach. In such cases it will be necessary to get them out of the system, for which purpose DR. HEARD'S WORM POWDERS are an excellent remedy. They should be given for 2 or 3 weeks after the discovery of the bots.

## THE ASCARIDE OR LARGE HORSEWORM.

This is seen in Fig. 85 and is usually from 4 to 10 inches long. These worms are sometimes very numerous in the horse and may give rise to frequent attacks of colic, diarrhea and debility. They also cause bad condition and predispose to disease. It has been estimated that the eggs produced by a single female often number several millions. They are usually taken into the alimentary canal with food and water. The treatment will consist in giving DR. HEARD's WORM POWDERS every day for ten days, to be followed by 6 drams of Barbadoes aloes and 1 dram of ginger made into a ball to be given on the tenth night after beginning the powders.

## OXYURIS, PIN-WORM, THREAD-WORM

This worm is very common in horses that have been pastured on grass in the summer, and is shown in Fig. 86.

These parasites often cause debility and loss of flesh, and cannot be got rid of too soon.

*Treatment.*—Give HEARD's WORM POWDERS as directed and on the 5th day after beginning the powders, inject into the rectum one ounce of turpentine mixed with a pint of linseed oil. On the 12th day give a ball containing 6 drams of Barbadoes

Fig. 85. The ascaride or large horse worm.

aloes and 1 dram of ginger after which the horse must be allowed to rest two or three days.

Fig. 86. Oxyuris or horse thread worm.

## DISTOMA HEPATICUM—LIVER-FLUKE.

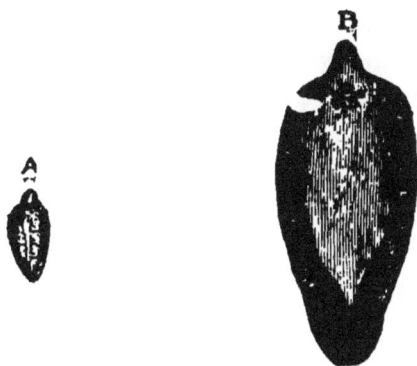

Fig. 87. Young flukeworm.          Fig. 88. Adult flukeworm.

The disease caused by this parasite is very common in some seasons and in some localities. It is peculiar to sheep and is known as liver-rot, fluke-disease, sheep rot, etc. In Figs.87 and 88 are seen specimens of the parasites with their size and shape. The body is flattened and the color is pale-brown. It is peculiar to low, wet pastures and wet seasons. Its life-history is very interesting. It passes a portion of its life in one of the snails. As these are found only in moist places, it explains the cause of the parasite being found only on wet pastures or during extraordinary wet seasons.

*Symptoms.*—For the first three months after sheep are infected, they will often fatten very fast. This is soon followed by diminished appetite and great thirst. Wasting now begins and continues until death occurs from anemia, unless the flukes are expelled and convalescence begins. The latter event usually occurs in May or June, but unfortunately most of the diseased sheep die before that time. On making an examination of the liver of a sheep that has died of this disease the flukes will be seen in the bile-ducts of the organ, varying in size from the young to the full-grown, as seen in Figs. 87 and 88. The losses of sheep from this disease in the whole world are said to be much greater than those from any other known disease. The losses in England in one year were estimated at 3,000,000 sheep.

*Treatment.*—Sheep should always be kept away from wet pastures. As to medicines, the infusion of pine-leaves, spruce-leaves, oak-leaves, walnut-leaves, lime-water, sulphate of iron, and common salt all have their special advocates. The two last seem to have the preference.

The experience of most observers has been that when

the parasite has once established a foothold in a sheep, it is almost useless to attempt to dislodge it.

## STRONGYLUS FILARIA, LUNG-WORM.

This worm, which is so destructive to young lambs and calves, is seen in Figs. 89 and 90. The disease which it produces is known as "hose in lambs," white-skin, pelt-rot, etc. When the parasites are present in large numbers, they cause bronchitis and pneumonia.

Figs. 89 & 90. Two specimens of strongylus filaria.

*Symptoms.*—Coughing attracts, early attention; if the animal is made to move, a fit of coughing is sure to occur. There is a thick discharge of mucus from the nostrils, and the worms are often found in it. As the disease advances the animal becomes greatly weakened and loses condition ; the breathing becomes labored, and all the symptoms of pneumonia may be present. The animal looks dull and stupid and does not care to move.

*Treatment.*—To prevent infection, sheep should be kept on new pastures. The old pastures on which infected sheep have been kept will contain the eggs of the worms, which will reach the lungs after being taken in with the food. The fumes of sulphur is about the best remedy after the disease has once got a foothold. The sheep must be placed in a closed building, and a small quantity of sulphur set on fire. This will produce hard fits of coughing, during which large numbers of the worms will be expelled. The animals must not remain in the sulphurous atmosphere more than ten minutes, as there is danger of suffocation. Another good remedy is the following : Oil of tar 1 oz., oil of turpentine 1 oz., sweet oil 6 ozs. Give 10 to 20 drops in the nostrils daily.

## CENURUS CEREBRALIS.

This worm produces the disease in sheep known as Gid, Turn-Sick, Hydatid on the Brain. The cenurus cerebralis is the cystic form of the tenia cenurus, which is one of the common tape-worms of the dog. In Fig. 91 is seen a cyst containing a number of the eggs of this worm. These cysts are found in the brain, and are the cause of the curious symptoms seen in gid. In Fig.

92 is seen the fully developed tape-worm as found in the dog. The eggs from the cyst are taken in by dogs

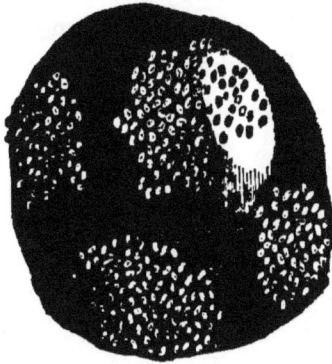

Fig. 91. Cyst containing eggs of cenurus cerebralis.

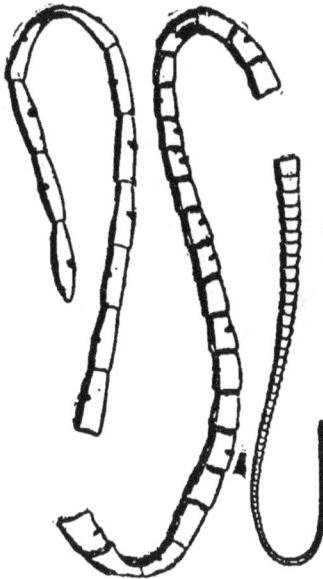

Fig. 92  Tenia cenurus, a dog tapeworm.

when devouring the head of the sheep, and the continuation of the development produces the tenia cenurus or dog tape-worm. As the worm developes, the eggs are voided by the dog. They are voided in great numbers and are distributed on the grass over which the dog has passed. The sheep eat them with the grass, and in this way the parasite gains access to the internal organs.

*Symptoms.*—In from 8 to 20 days after the parasite has reached the brain, signs of congestion appear ; the head is carried lower or higher than usual and to one side ; the animal performs forward movements to the right or left, in a circle, or it rotates in one spot—going round as if on a pivot. In some cases the animal acts as if intoxicated, and frequently stumbles and falls. It will keep by itself and shun other sheep. In some cases it goes straight forward, lifting the head and feet higher than usual. The symptoms will vary according to the situation in the brain that the parasite occupies; we may have partial or complete paralysis.

*Treatment*—We shall not recommend any treatment for cases in which gid has already developed, but must be content to advise that the heads of sheep that have died from it should be burnt ; and in fact the heads of sheep should never be fed to dogs. If this rule was carefully observed, the species of cenurus cerebralis would soon become extinct.

## THE ADMINISTRATION OF MEDICINES.

There are various methods of administering medicines to the domestic animals, depending on the nature of the medicine to be given and the species of animal to which it is proposed to give it.

Liquid medicines are given to the horse mostly by the mouth, sometimes by the rectum and in some cases by injecting it under the skin. In Fig. 93 is seen a

Fig. 93. Administering a drench.

picture of a man in the act of giving a dose of liquid medicine by the mouth. It will be seen that one man is holding up the head of the animal by placing a loop of rope or a leather strap in the mouth and then inserting one of the prongs of a fork and holding up the mouth to the height desired ; the other man is seen standing on a box holding a bottle so that the contents will slowly flow into the mouth of the horse to be drenched. One or two warnings must be given here. Don't drench the horse through the nostrils. If the horse gives a cough while being drenched, lower the head immediately. Not more than 2 fluid ounces should be placed in the mouth at once and see that the animal swallows that before getting another two ounces. The throat should not be squeezed to make the animal swallow. Don't hold the tongue while drenching. Don't be in too much hurry to get the bottle or drenching horn emptied. This is a very simple way of giving bulky medicines that are in a liquid state and the most of our best medicines for the common diseases such as colic, chills and indigestion can only be given in this form. Oatmeal gruel and linseed tea or milk and eggs for nutriment can often be advantageously given by this method. In lock-jaw or other diseases where drenching is not practicable the medicines may be given by being injected into the rectum with the syringe seen in Fig. 94. Cows may be drenched by having an

Fig. 94. Syringe for rectal injections,

assistant steady the head by holding the horn and nostrils and using a bottle or drenching horn.

Dogs can be made to take liquid medicine by simply holding the mouth open and allowing it to drop downward toward the gullet.

Whenever the properties of the medicines will allow it, for dogs and horses their administration in pill form will be most desirable for many reasons. The reason that this is not more commonly done in the case of horses is the fear of having the hand crushed. This can be very readily overcame by the use of the little instrument seen in Fig. 95 called a " mouth speculum,"

Fig. 95. Mouth speculum.

which is inexpensive and will last a lifetime. This should be placed in the mouth, the tongue held with the left hand and the ball between the thumb and first and second fingers of the right hand ; this should now be passed back through the round opening in the mouth speculum toward the throat, going right back over the tongue and leaving the ball as near the throat as possible. After trying this two or three times anyone may become proficient and rarely make a mistake. To give pills to dogs the mouth should be held well open and the pill dropped as far back towards the throat as possible and the mouth closed for a minute or two.

# TABLE OF DISEASES AND THEIR REMEDIES.

**Abortion.**—Isolate aborting from pregnant animals. Burn or bury fetus and afterbirth.

**Abscess.**—Fomentations of hot water; poultices; HEARD'S AMERICAN EMBROCATION; use the knife as early as possible to evacuate the pus after which inject and dress with HEARD'S HEALING LOTION.

**Actinomycosis.**—Fatten the animal and send to the butcher as early as possible, but destroy all diseased parts.

**Anemia.**—Tonics; iron and arsenic, gentian, quinine, HEARD'S CONDITION POWDERS, good air; clip long coated horses; phosphate of lime in growing animals. Two ounces of linseed oil a day until the bowels are slightly relaxed.

**Anthrax.**—Inoculation with attenuated virus as a preventive. Burn or bury deeply all animals that die of the disease and remove all young stock from the same pasture. Disinfect freely with corrosive sublimate.

**Aphtha—Thrush—Vesicles in mouth.**—Wash out mouth three times a day with a solution of borax, a teaspoonful in a half pint of water. Give lime water or solution of bicarbonate of soda to drink. Mix a little linseed with bran mashes for food.

**Apoplexy parturient**; see **Milk Fever.**

**Arthritis—Inflammation of joints—Synovitis.**—
Hot fomentations, poultices mixed with HEARD'S AMERICAN
EMBROCATION, perfect rest; HEARD'S AMERICAN EMBROCA-
TION well rubbed in and a bandage applied to cause a
blister. Place horse in slings if very lame. Internally:
cathartic ball ; if pain is very great a half a bottle of
HEARD'S MAGIC MIXTURE twice a day; chloral.

**Ascarides ;** see **Worms.**

**Asthma** in horses; see **broken wind.**

**Azoturia.**—Aloes cathartic ball, frequent injections
of warm water. If very much excited give a bottle of
HEARD'S MAGIC MIXTURE. Saltpetre in large doses three
times a day. Remove urine with catheter, apply HEARD'S
AMERICAN EMBROCATION to loins. Place animal in slings.
Give very little food for two or three days.

**Black-leg.**—Prevent by keeping young cattle and
sheep out of infected pastures. Burn or deeply bury
all animals that die of this disease.

**Bladder irritable.**—Linseed tea, change diet ; bicarbonate of soda, iodide of potassium, sulphate of iron.

**Bog spavin.**—Rest ; HEARD'S AMERICAN EMBROCATION;
high heeled shoe; hot fomentations if very lame. Truss
in young animals.

**Boils** see **Abscess.**

**Bots.**—HEARD'S WORM POWDERS as directed on the
box.

**Broken wind—**HEARD'S CONDITION POWDERS; give
only a small quantity of hay; allow only small quantity
of water immediately before hard work.

**Bronchitis.**—Keep surface of body warm with extra
clothing and bandages. HEARD'D AMERICAN EMBROCATION

to front and sides of chest three times a day. Give following ball; quinine 1 dram, nitrate of potash 4 drams, tartar emetic 1 dram, to be repeated three times a day until the fever has disappeared.

**Bruises.**—Hot water fomentations, the application of HEARD'S AMERICAN EMBROCATION.

**Burns and Scalds.**—HEARD'S HEALING LOTION ; Vaseline ; Lime water.

**Capped Hock.**—Prevent kicking ; HEARD'S AMERICAN EMBROCATION; use high-heeled shoe ; ointment of biniodide of mercury.

**Capped Elbow.**—If large and recent evacuate fluid from bottom surface and keep open until healed inside, dress twice a day by injecting a little of HEARD'S HEALING LOTION. The application of HEARD'S AMERICAN EMBROCATION cures slight cases. Prevent the horse from lying down by securing to a high ring in the front of the stall until cured.

**Caries.**—Remove diseased bone, and dress with HEARD'S HEALING LOTION twice a day.

**Cataract.**—Incurable except by difficult operation.

**Cerebro-spinal Meningitis.**—see Spinal Meningitis

**Choking.**—External pressure over the obstruction ; give linseed oil ; use new rope or probang.

**Colic Flatulent.**—HEARD'S MAGIC MIXTURE ; Injections of warm water and soap ; tapping abdominal contents with trochar ; Hot fomentations long continued to abdomen, bicarbonate of soda in solution.

**Colic—Spasmodic.**—HEARD'S MAGIC MIXTURE, injections of warm water and soap ; hot fomentations to abdomen, application of HEARD'S AMERICAN EMBROCATION, to abdomen.

**Constipation.**—Aloes Cathartic ball ; Linseed oil ; Injections of warm water and soap. Give soft food.

**Corns in feet.**—Apply bar shoe ; soak feet in soaking tub of hot water for 4 hours a day ; blisters to coronet ; apply Heard's Hoof Liniment to hoof to get the horn healthy.

**Cough.**—HEARD'S CONDITION POWDERS ; restrict the quantity of hay to less than 8 lbs. a day. Baking soda in the drinking water. Clip horses with long coats.

**Crib-biting.**—Apply strap to neck ; use iron stall fittings; apply muzzle.

**Curb.**—HEARD'S AMERICAN EMBROCATION, firing and blistering ; high heeled shoe ; rest while lame.

**Debility.**—HEARD'S TONIC CONDITION POWDERS: Sulphate of iron ; Fowlers' solution, quinine, generous diet that is easily digested, as milk, oatmeal drinks and linseed tea.

**Diabetes.**—Polyurea: Bicarbonate of soda; iron and iodine. Change food.

**Diarrhea.**—Bicarbonate of soda; chalk; gum catechu; laudanum; restrict the quantity of water, when obstinate, try beef tea, milk and eggs.

**Dislocations.**—Replace bones in their natural position. Retain in place by splints, bandages and blisters. Reduce surrounding inflammation by hot fomentations and HEARD'S AMERICAN EMBROCATION. Sling if necessary.

**Distemper.**—See **Influenza.**

**Dropsy.**—Nitrate of potash in large doses, hot fomentations, and scarifications with a lancet. Allow very little food until recovery is advanced. As soon as kidneys act briskly give iron and the vegetable tonics : HEARD'S CONDITION POWDERS.

**Eczema.**—HEARD'S DERMAL LINIMENT; Extract of witch-hazel; solution of borax; infusion of arnica root; glycerine, Internally, HEARD'S CONDITION POWDERS.

**Emphysema.**—This is a swelling filled with a gas. Puncture, hot fomentations, HEARD'S AMERICAN EMBROCATION.

**Emphysema of Lungs.**—See **Broken Wind.**

**Epizootic.**—See **Influenza.**

**Erysipelas.**—Use HEARD'S HEALING LOTION locally and poultices where practicable, also hot fomentations. Internally, stimulants and tonics.

**Farcy**—See **Glanders.**

**Fever.**—Quinine in large doses ; nitrate of potash, alcohol ; keep on very low diet until the high temperature has disappeared.

**Fistula.**—Lay open sinus with knife or iron heated to a white heat ; insert seton, dress with HEARD'S HEALING LOTION ; prevent the animal from rubbing or otherwise irritating it. Inject strong solution of corrosive sublimate, when obstinate, slough out walls of sinus with plug of corrosive sublimate.

**Fleas.**—HEARD'S MANGE CURE; Persian insect powder, pine saw-dust bed for dogs.

**Foot rot in Sheep.**—Remove loose horn ; powdered sulphate of zinc dressings, strong bichloride of mercury solution, dress feet at least three times a week.

**Founder.**—See **Laminitis.**

**Fractures.**—See **Index.**

**Frost bites.**—See **Scratches.**

**Garget.**—See **Mammitis.**

**Glanders and Farcy.**—Destroy animal and isolate all suspected cases, and report to local health authorities. This disease is also dangerous to man.

**Hematuria.**—Salts of iron, iodide of potassium, bicarbonate of soda, arsenic, injections of warm water and soap, HEARD'S CONDITION POWDERS, feed on good nutritious diet.

**Hemorrhage.**—Apply bandages where practicable, oakum soaked in strong bichloride of mercury solutions, extract of witch-hazel, opium, continuous pressure in the neighborhood of the wound. Actual cautery.

**Hernia.**—Treated by operations and bandages where recent.

**Hydrophobia.**—See **Rabies.**

**Indigestion.**—Soda bicarbonate, iodide of potassium tartar emetic, calomel, injections of warm water and soap. HEARD'S CONDITION POWDERS. If pain is present give HEARD'S MAGIC MIXTURE.

**Inflammation.**—Internally. Small doses of aconite, in the early stage, bleeding in acute cases, belladonna, tartar emetic, cathartics, nitrate of potash.

Externally. Hot fomentations followed by HEARD'S AMERICAN EMBROCATION.

**Influenza.**—Internal; quinine, nitrate of potash, antipyrine, acetanelid, chlorate of potash, whiskey; external, HEARD'S AMERICAN EMBROCATION to the throat, plenty of warm clothing and bandages to the legs. Diet; good nourishing food as milk, linseed tea, small quantity of oats when the temperature becomes nearly normal.

**Laminitis.**—Internally ; Barbadoes aloes, nitrate of potash, tartar emetic, salicylate of soda. Externally Remove shoes, soak feet in a soaking tub full of hot water, apply poultices to feet; if chronic, blister coronet, subsequently apply bar shoe.

**Lampas.**—Scarify with lancet, apply astringent lotion.

**Laryngitis.**—See **Sore throat.**

**Lice.**—HEARD's MANGE CURE, tobacco water, cleanliness.

**Lock-jaw.**—Large doses of cathartics and nitrate of potash, chloral, keep patient perfectly quiet, blisters to spine.

**Lung fever.**—See **Pneumonia.**

**Lymphangitis.**—Give cathartic ball, chlorate of potash, bicarbonate of soda. Externally, hot fomentations followed by HEARD's AMERICAN EMBROCATION. After two or three days give gentle exercise and HEARD's CONDITION POWDERS.

**Mammitis.**—Hot fomentations and poultices to udder, followed by HEARD's AMERICAN EMBROCATION, milk 3 or 4 times a day. Internally; cathartic medicine. Give very little food in the active stages of the disease.

**Mange**—Wash and apply HEARD's MANGE CURE as directed. Internally. HEARD's CONDITION POWDERS.

**Milk Fever.**—Bleeding, large dose of epsom salts, ergot, cold pack to spine, aconite, belladonna, chloral, aloes, spirits nitre, whiskey.

**Moon Blindness.**—Cathartic medicine, chloride of ammonium, calomel. Application to eye of HEARD's MOON EYE LOTION.

**Mud Fever.**—Internal. Cathartic medicine; bicarbonate of soda; followed by HEARD's CONDITION POWDERS.

External.   HEARD'S HEALING LOTION; poultices of linseed meal.

**Navicular Disease.**—Remove shoes; rest; soaking feet in cold water; cold water swabs: blisters to coronet; neurotomy.

**Necrosis.**—Remove dead bone; application of HEARD'S HEALING LOTION.

**Open Joints.**—Internal. Cathartics and nitrate of potash, tartar emetic. External, rest; hot fomentations and poultices and antiseptic injections, cold water irrigation; blister.

**Ophthalmia.**—See **Moon-blindness.**

**Over-reach.**—Hot fomentations; poultices; HEARD'S AMERICAN EMBROCATION.

**Paralysis.**—Strychnia; potassium iodide; sulphate of iron; friction to the spine and an application of HEARD'S AMERICAN EMBROCATION; easily digested food.

**Palpitation of Heart.**—Belladonna; aconite; digitalis.

**Peritonitis.**—Local hot fomentation; poultices: application of HEARD'S AMERICAN EBROCATION. Internally; aconite; chloral; opium; colomel; digitalis; injections of warm water in the rectum.

**Pharyngitis.**—See **Sore Throat.**

**Periostitis.**—Rest; hot fomentations and the application of HEARD'S AMERICAN EMBROCATION; cathartic ball.

**Pleurisy.**—Internal. Calomel; tartar emetic; nitrate of potash; digitalis; nux vomica; bleeding. Externally; applications to the sides of mustard made into a paste by the addition of HEARD'S AMERICAN EMROCATION; poultices to chest; plenty of surface clothing.

**Pneumonia.**—Quinine; whiskey; calomel, nitrate of

potash ; linseed oil in 4 oz. doses if the bowels are costive; camphor, plenty of drinking water ; if much debility, give milk diet ; enemas of warm water 3 times a day; the application of HEARD'S AMERICAN EMBROCATION 3 times a day to the chest and sides. HEARD'S CONDITION POWDERS to build up the system when recovering.

**Pumiced foot.**—see page 141.

**Purpura Hemorrhagica.**—Nitrate of Potash; sulphate of iron; sulphate of soda; calomel; ergot; spirits of turpentine; whiskey; application of hot fomentations; diluted embrocation; plenty of fresh air; milk diet; injections of hot water if bowels are costive.

**Quittor.**—Lay sinus open and allow perfectly free drainage; inject carbolic acid solution twice daily.

**Rabies.**—Pasteur inoculations with attenuated virus as a preventive.

**Rheumatism.**—Internally,Salicylate of soda; calomel; tartar emetic; colchicum; nitrate of potash; bicarbonate of potash ; antipyrin ; when chronic ; quinine; arsenic ; HEARD'S CONDITION POWDERS. Externally, HEARD'S AMERICAN EMBROCATION; hot fomentations; poultices; blisters.

**Ringbone.**—In early stages HEARD'S AMERICAN EMBROCATION; firing and blistering.

**Ringworm.**—Disinfect brushes, clothing, harness, and articles that come in contact with the parts affected. HEARD'S MANGE CURE ; tincture of iodine.

**Roaring.**—To relieve distressed breathing after violent exercise give HEARD'S CONDITION POWDERS.

**Spinal Meningitis.**—Place in slings; linseed oil; hypo-sulphite of soda; nitrate of potash; muriate of ammonia; injections; mustard blister to spine.

**Saddle Galls.**—Apply properly fitting saddles and if the skin is broken HEARD'S HEALING LOTION. If skin is swollen and inflamed and not broken apply diluted Embrocation one part to 4 of water.

**Scab in Sheep.**—Blue Mercurial ointment or the sheep dip recommended in the article on Scab.

**Side bone.**—Rest; fire aud blister.

' **Sore shins.**—Rest; hot fomentations; HEARD's AMERICAN EMBROCATION. Cathartic ball.

**Sore throat.**—Poultices to throat; HEARD's AMERICAN EMBROCATION; nitrate of potash; camphor; swab inside of throat with solution of chlorate of potash. Tar and licorice paste on tongue.

**Spavin-bone.**—Fire and blister.

**Spavin-bog.**—HEARD's AMERICAN EMBROCATION.

**Splint.**—When causing lameness, fire and blister and rest.

**Speedycut.**—Hot fomentations; HEARD's AMERICAN EMBROCATION; open abscesses if they are formed; lessen the quantity of work.

**Sprains.**—Hot fomentations; HEARD's AMERICAN EMBROCATION ; if back tendons are sprained, use shoe with very high heel calks.

**Staggers.**—Bleeding; cathartic ball; HEARD's CONDITION POWDERS.

**Strangles.**—External. Poultices to throat and swellings; HEARD's AMERICAN EMBROCATION; open abscess when ripe. Internal same as for Influenza.

**Synovitis.**—See **Open Joint.**

**Tetanus.**—See **Lock Jaw.**

**Thoroughpin.**—DR. HEARD's AMERICAN EMBROCATION; firing and blistering,

**Thrush in Feet.**—HEARD's HEALING POWDERS. Cathartic ball.

**Udder.**—Inflammation of. Hot fomentations; HEARD'S AMERICAN EMBROCATION; if abscess forms it must be opened.

**Uterus.**—Inflammation of. Woolen clothes wrung out in hot water, and placed under abdomen and over the loins followed by application of HEARD'S AMERICAN EMBROCATION. Cathartic medicine.

**Warts.**—Remove with knife or ligature.

**Windgalls.**—HEARD'S AMERICAN EMBROCATION and bandages.

**Worms.**—DR. HEARD'S WORM POWDERS.

**Wounds.**—HEARD'S HEALING LOTION; bandages; and touch proud flesh with sulphate of zinc.

# TABLE OF MEDICINES AND THEIR DOSES.

The doses given are for an ordinary sized horse of three years old and upwards. Under three years, the dose will have to be lessened according to the age, decreasing the dose so that from 1½ to 3 years, the quantity should be about one-half; from 9 to 18 months, ¼ the dose mentioned.

Acacia gum, 2 to 6 oz.
Acetanilid, ½ to 2 drams.
Acid Acetic (diluted) 1 to 3 oz.
Acid Arsenious (Arsenic) 2 to 8 grs.
Acid Boracic, 1 to 3 dr.
Acid Carbolic, ½ to 1½ dr.
Acid Hydrochloric, (dil.) 1 to 3 dr.
Acid Tannic ½ to 1½ dr.
Aconite (Tincture) 20 drops to 1 dr.
Aconitine ¼ to 1 gr.
Aether (Sulphuric) 1 to 1½ oz.
Aether Nitrous (Spirits of Nitre. ½ to 3 ozs.
Alcohol 1 to 3 ozs.
Aloes (barbadoes) 4 to 8 drs.
Aloin ½ to 2 dr.
Alum Sulphate 2 to 5 drs.
Ammonia: Aromatic spirits ½ to 1½ ozs.
Ammonia Chloride 2 to 6 drs.
Anise Seed ½ to 1½ ozs.
Antimony: Black 1 to 3 drs.
Antipyrine ½ to 3 drs.
Antifebrine ½ to 3 drs.
Areca Nut ½ to 1½ ozs.
Arsenic 2 to 8 grs.
Atropine sulphate ½ to 2 grs.
Belladonna (extract) ½ to 1½ drs.
Borax 2 to 6 drs.
Bicarbonate of Soda 1 to 8 drs.
Calomel ½ to 4 drs.
Camphor 1 to 4 drs.
Cantharides 2 to 15 grs.
Capsicum 10 to 30 grs.
Caraway seeds 2 to 8 drs.
Carbolic acid ½ to 1½ drs.
Carbonate of Ammonia ½ to 3 drs.
Carbonate of Iron 1 to 4 drs.
Carbonate of Magnesia 1 to 6 drs.
Cascara Segrada 20 to 40 grs.
Catechu 1 to 8 drs.
Chalk 2 to 8 drs.
Chloral Hydrate ½ to 2 ozs.
Chloroform ½ to 1 oz.
Chlorate of Potash ½ to 1 oz.
Colchicum ½ to 3 drs.
Cod Liver Oil 1 to 4 ozs.
Croton Oil 5 to 20 drops.
Copper Sulphate ½ to 2 drs.
Cream of Tartar 1 to 4 ozs.
Creolin 1 to 4 drs.
Digitalin ¼ to 1 gr.
Digitalis ½ to 2 drs.
Elaterium 10 to 30 grs.

Ergot ¼ to 1 oz.
Eserine Sulphate ¼ to 3 grs.
Ether Sulphuric ¼ to 1 oz.
Fenegeric ¼ to 1 oz.
Fowlers Solution ¼ to 1 oz.
Gentian 2 to 8 drs.
Ginger 1 to 4 drs.
Glycerine 1 to 4 ozs.
Hyoscymus Extract 1½ to 4 drs.
Hyoscymine 1 to 3 grs.
Iodine 10 to 30 grs.
Iodide of Iron 1 to 3 drs.
Iodide of Potash ¼ to 2 drs.
Ipecacuanha 1 to 2 drs.
Jaborandi ¼ to 1 oz.
Jalap ¼ to 2 ozs.
Kino 1 to 3 drs.
Linseed Oil 4 oz to 1 quart.
Liquor Arsenicalis ¼ to 1 oz.
Magnesia Sulphate ¼ to ½ lb.
Morphine 3 to 10 grs.
Nicotine 2 to 6 grs.
Nitre (Saltpetre) ½ to 2 ozs.
Nux Vomica 20 to 60 grs.
Opium powdered 1 to 2 drs.

Pepsin ¼ to 2 drs.
Phosphorous ¼ to 2 grs.
Phenacetin 2 to 4 drs.
Peppermint Oil 15 to 30 drops.
Pilocarpine 2 to 5 grs.
Podophylin 10 to 30 grs.
Potash Nitrate (Saltpetre) ½ to 2 ozs.
Potash Bromide of 2 to 6 drs.
Potash Iodide ¼ to 2 drs.
Prussic Acid (diluted) ¼ to 1 dr.
Quinine ½ to 3 drs.
Resin ¼ to 1 oz.
Rhubarb ¼ to 1 oz.
Santonin 10 to 40 grs.
Salicylic Acid 1 to 3 drs.
Salicylate of soda 1 to 4 drs.
Strychnia ¼ to 3 grs.
Tartar Emetic ¼ to 6 drs.
Turpentine Spirits ½ to 2 ozs.
Tincture of Aconite 20 drops to 1 dr.
Tincture of Opium Laudanum) ¼
   to 2 ozs.
Tincture of Belladonna ¼ to 2ozs.
Valerian Root 2 to 12 drs.

14

D R. HEARD'S preparations all keep perfectly good for any length of time.

All questions relating to the use of these medicines will be cheerfully answered by mail. In fact, Dr. Heard wishes his patrons to communicate with him whenever they are in doubt about the treatment of any given case.

If your Harness or Drug store does not keep a supply on hand it is because they wish to run off an inferior preparation on which they reap a larger profit, but remember that it always pays the consumer to buy the best.

Dr. Heard will ship his preparations to any address on receipt of the advertised price, with a liberal discount on large orders.

Local Agents wanted everywhere for Dr. Heard's Books and Veterinary Preparations.

# INDEX.

# HOW TO OBTAIN ADVICE ON ANY DISEASE WHERE THE SERVICES OF A VETERINARIAN ARE DIFFICULT TO OBTAIN.

If the animal is sick give the details of symptoms in answer to the following questions :

What species of animal ?

Ma`e or female ?

Ag⟨ ?

How long sick ?

Is appetite good or bad ?

What kind of labor has it been used at ?

How many respirations (times of breathing) in a minute ?

Is there any cough and if so how long have you noticed it ?

Does the animal drink the average quantity of water ?

Are the bowels working all right ?

If you have a thermometer take the temperature of the patient and report the same.

Are there any swellings on any part of the animal, and if so, where ?

Have you examined the mouth and found anything wrong with the teeth ?

How has the animal been fed for the past three months ? and describe fully any other symptoms that may be noticeable, and enclose one dollar addressed to

**DR. J. M. HEARD, M.R.C.V.S.**

**119 West 56th St., New York City,**

**N. Y.**

# TO GET ADVICE ON LAMENESS SEND FULL ANSWERS TO THE FOLLOWING QUESTIONS.

Age of horse?

How long lame and in which limb?

Do you notice any swelling, and if so where?

Did you first notice the animal lame while at work or when being taken out of the stable?

Does the animal get better or worse when exercised?

Is it more noticeable when going up or down hill?

Is there any tenderness on pressing any part of the limb, and if so, at what part?

How has the horse been managed since you first noticed the lameness?

Has any treatment been applied to get rid of the disease and if so, what?

**By forwarding answers to the above questions, and enclosing One Dollar addressed to**

## DR. J. M. HEARD, M.R.C.V.S.

**119 West 56th St., New York City.**

**N. Y.**

**Advice as to the best method of treatment will be forwarded by return mail.**

Syringe.

Trochar.

Thermometer.

Tooth Rasp.

Mouth Speculum,

# INSTRUMENTS THAT ARE ABSOLUTELY NECESSARY TO ALL STOCK OWNERS.

One of the most serviceable of all instruments is the **Syringe** seen on the opposite page ; it is in frequent demand for giving injections of warm water and soap in a great many diseases. The price of this syringe, which holds 36 ounces is $3.50.

**Thermometer.**—This is a very important instrument for taking the animal's temperature and is a great aid in making a diagnosis. Before using, see that the top of the column of mercury is below the figure 100. Now place thermometer in the rectum, bulb end first, and allow it to remain in four minutes, remove the thermometer and observe the height of the column of mercury and that will indicate the temperature of the animal. The price of a good self-registering thermometer is $1.50. The cheaper ones all have some defect.

**Trochar.**—The use of this instrument is explained on pages 118 and 119. The price is $2.25.

The **Mouth Speculum** is used to keep the mouth open for the purpose of passing back the hand towards the throat to give balls to horses or to allow of a ready examination of the back teeth. The price is $1.00.

The **Tooth Rasp** is the instrument used for filing off the sharp edges or points which frequently project from the molar or back teeth as explained on page 167. The price is $2.25 and 25 cents each for new file blades to replace the file when it is worn out. By sending the price of any of those instruments either together or separately with an order describing which instruments are required they will be immediately forwarded to the nearest express office. The thermometer and trochar may be sent by mail if requested when six cents in stamps should be added to the price as given above.

The five instruments will be forwarded to any address on receipt of $10.00.

### Address all communications to

**DR. J. M. HEARD, 119 West 56th St.,**

**NEW YORK CITY.**

## DR. HEARD'S

# Magic Colic Mixture

## For Horses and Cattle.

This mixture is unequalled as a remedy for Colic, Chills, Inflammation of the bowels, Gripes, Indigestion, Wind Colic, Staggers, all Liver and Stomach complaints, Kidney and Bladder diseases and all Bowel derangements.

No stock owner should be without a supply of this medicine as it has saved the lives of thousands of animals and keeps its strength for any length of time.

Price one Dollar per bottle and may be obtained at all Harness and Drug stores.

Will be forwarded to any address in the United States and Canada on receipt of price.

Address

## DR. J. M. HEARD, M. R. C. V. S.

119 WEST 56TH STREET,

NEW YORK CITY, N. Y.

# DR J. M. HEARD'S

# WORM POWDERS

These worm Powders have been successfully used for many years past as a safe and sure remedy for all kinds of Worms especially those described on pages 185-186-187 in this book.

Price one dollar per box.

For sale by all Harness and Drug stores and sent free on receipt of price to any address in the United States and Canada.

Address

## DR. J. M. HEARD, M. R. C. V. S.

### 119 WEST 56TH STREET,

### NEW YORK CITY, N. Y.

When ordering from Harness or Drug stores either wholesale or retail, be sure and specify by name Dr. Heard's preparations to prevent the sending by them of inferior imitations of Dr. Heard's medicines.

DR. HEARD'S

# Chart for Horse Owners.

Printed on stout board 16 x 20 inches, containing 26 engravings with accompanying explanatory text, arranged to be hung in offices, making an artistic and instructive decoration. Will be mailed inside a suitable mailing tube on receipt of an application accompanied by Ten cents in stamps.

---

# HORSE SHOEING, PAST & PRESENT.

In preparation and will be issued about March 1st., 1894. A revised and enlarged edition of "HORSE SHOEING, PAST AND PRESENT," will be well illustrated and bound in cloth.

Price Fifty Cents.

Sent free by mail on receipt of price, when issued.

Address

## DR. J. M. HEARD, M. R. C. V. S.

119 WEST 56TH STREET,

NEW YORK CITY, N. Y.

# DR J. M. HEARD'S
# Chart for Drug Stores.

This chart is gotten up especially for Druggists, and contains a Table of Animal Diseases and their remedies; with a Table of Medicines and their doses for the Horse, Ox and Dog.

Printed on stout cardboard suitable for hanging in Drug Laboratories, where it will be a valuable reference chart, enabling the owner to correctly prescribe for common animal ailments to his own profit as well as that of his patrons.

Sent free to all Drug Stores on application to

# DR. J. M. HEARD, M.R.C.V.S.
## 119 WEST 56TH STREET,
## NEW YORK CITY, N. Y.

---

# How to Tell the Age of the Horse

### By DR. J. M. HEARD, M. R. C. V. S.

ILLUSTRATED.                    PRICE, 30 .

——SOLD BY——

## M. T. RICHARDSON,
### 86 READE STREET, NEW YORK CITY.

# · TESTIMONIALS ·

THE following are a few of the numerous letters that have been received within the past few months:

"One of the most troublesome things we have had to contend with for several years has been the bunches that grow on the limbs of horses as the result of bruising with the opposite foot. These bunches often become hard and colloused and no remedy that we could find or that the veterinarians could suggest seemed to be of any benefit until we accidentally came across your EMBROCATION, since which we have not had a case of this kind that has not been completely cured by the application of your EMBROCATION as directed. I can therefore recommend it in the highest degree for all purposes where it is necessary to get rid of either temporary or chronic enlargements."

CHAS. W. DICKEL, of Dickel's Riding Academy,
N. Y. City.

"I have used your EMBROCATION with great success. One case in particular I wish to mention. A horse owned by Mr. Ginter the large cigarette manufacturer, was badly affected with eczema; he broke out all over in large abscesses and in two days the EMBROCATION entirely killed the whole thing, and in two weeks the hair had grown out again so that no one could see that there had been anything the matter with him. I would not be without it at any cost."

HERBERT CODD,
Richmond, Va.

"We have used large quantities of your MAGIC COLIC MIXTURE for all kinds of painful diseases of the bowels and the various forms of colic which are continually occurring in some of the two thousand horses of our establishment and I have always found it a very efficient remedy.

I. HOUGH, V. S.
3d Ave. R. R. Stables,
New York City.

"We have used a number of bottles of your AMERICAN EMBROCATION on horses with sore throats and various forms of lameness, and can affirm, without hesitation, that it is the best liniment we have used in thirty years experience. Please send us one dozen as soon as possible, as we cannot do without it."

J. H. WIKE,
Supt. of Central Park Riding Academy Stables,
New York City.

"I have had several cases of colic in horses under my charge in the last 2 years and never failed to cure it in a very short time with a bottle of your MAGIC MIXTURE."

T. J. ROEBUCK,
West Townsend, Mass.

"I have used your AMERICAN EMBROCATION on several bad cases and can honestly say it is the best I ever used, and it will surely do everything you claim for it."

E. T. LANE,
Supt. of Brooklyn Riding and Driving Club,
Brooklyn, N. Y.

"One of Mr. F. S. Kinney's horses was last week attacked with a violent colic and we administered a bottle of your MAGIC COLIC MIXTURE which brought entire relief in a few minutes, in fact its action was so prompt that I would advise every horse owner to keep some of the MIXTURE continually on hand for sudden emergencies."

W. H. MITCHELL, Supt.,
Kinnelon, Morris Co., N. J.

"I have used your HOOF LINIMENT and MOON EYE LOTION for some years past and they are the best preparations that I have ever known to be used on horses."

MICHAEL CLANCY,
Philadelphia, Pa.

"I have used your DERMAL LINIMENT on all our carriage horses to rub down with after coming in from work every day during the past summer and I have not had a shake or a chill or a particle of soreness or lameness during all the time that it has been used. It is also a good cleansing agent and makes a great improvement in the appearance of the coat."

F. COOK, Bellevue Court,
Newport, R. I.

"It gives me great pleasure to testify to the excellence of your EMBROCATION. With such a large number of horses as I have under my charge it is in constant use, and believe me, I would not be without it if its cost were double what it is, and I should still esteem it the cheapest liniment in the market."

WILLIAM MATHEWS,
New York City.

"I have used the EMBROCATION in a bad case of chronic sprain and a capped hock, and I never saw anything act so promptly in relieving lameness and swelling as this EMBROCATION, and I shall certainly recommend it to all my friends who have the care of valuable trotters; in fact, I wouldn't be without it if the cost was three times as high as it is.

WILLIAM CURLEY,
Supt. Frank Work's Stable,
New York City.

"I have used your HEALING LOTION for the healing of strangles sores and am surprised how quick they heal up."

MAURICE McAULIFFE,
Druid Hill Ave.,
Baltimore, Md.

"I have used the EMBROCATION received from you on several cases of long standing lameness. I can cheerfully testify to the great benefit that our horses have received from its use, and have found it especially beneficial in shoulder lameness."

J. BAUGHAN.
Supt. 26th Street Stables,
New York Transfer Co.

"About two months ago I had to have a large tumor removed from a horse's shoulder and after the tumor was cut out there was a hole in the flesh that I could easily lay my fist in and I expected it would take at least three months to heal it, but by the use, twice a day, of your HEALING LOTION it was entirely closed up in a month's time."

THOS. CONATY,
E. 52nd Street, New York City.

"Last summer one of my horses had a large abscess on the thigh, which, when opened, was relieved of over a quart of fluid. I used your HEALING LOTION as directed and the wound healed up very quickly without leaving any swelling or other bad effects. It certainly made a wonderful cure."

W. H. ADAMS.
Weaver Ave.,
Newport, R. I.

"I have used several dozen bottles of your EMBROCA-TION, and always with the best success."

J. PARSONS,
New York City.

"I have used your EMBROCATION for sore throats for the past 6 years and always with success."

FRANK DUFFY,
German Street,
Baltimore, Md.

"I have used your EMBROCATION on several horses with curb and found that it takes them off clean in every case, and I recommend it as being the most valuable LINIMENT in the market for horse dealers."

H. W. SWAN,
Bourn Street,
Providence, R. I.

"In the last month two of our horses have had bad attacks of colic from over driving and standing in the cold afterwards, and in each case immediate relief from the pain followed the giving of a bottle of your MAGIC MIX-

TURE. It is the greatest remedy for colic that we have ever used, and everybody who owns a horse should always have a bottle on hand."

BROWN & EVANS,
6th Ave., New York City.

"Having used your EMBROCATION for pinkeye, distemper, and sore throats in green horses for several years past, it gives me pleasure to testify to its very beneficial effects in all those diseases. I also use it continually for sprains and bruises, and would not be without it for a good deal."

STRAUS & IMMEN,
159 and 161 E. 24th St., N. Y.

"I have used the two dozen bottles of EMBROCATION with every satisfaction. In cases of splint in their infancy I think it superior to any imported EMBROCATION."

HERBERT CODD,
Richmond, Va.

"I have used your HOOF LINIMENT for the past four years, and have always found that its use keeps the horn naturally moist and tough and makes a growth of strong horn at the coronet. It is far ahead of anything that I have ever seen used."

JAMES SKELLY,
40 Tennison Street,
Boston, Mass.

"Your CONDITION POWDERS I have always found to produce a fine coat and create a good appetite."

GEORGE DAWSON,
St. Paul Street,
Baltimore, Md.

"I have found your CONDITION POWDERS very useful as producers of good appetite after any weakening disease."

WILLIAM GILES,
Clinton Street,
Buffalo, N. Y.

"A couple of weeks ago one of our horses got kicked in the hip and it made quite a large deep wound in through the skin and muscles. I used your HEALING LOTION on it twice a day, as directed, and now it has all healed up. I have often seen similar cases take two months to heal."

JOHN CASSON,
W. 52nd Street, New York City.

" The WORM POWDERS I received lately have fully accomplished the extinction of the worms with which my horses were affected."

ABNER BAKER,
Fall River, Mass.

"The HOOF LINIMENT that I ordered from you has grown down such a strong hoof that there is now no sign of the quarter crack that I referred to in my letter."

THOS. WILLIS,
Lenox, Mass.

"I have used large quantities of your HOOF LINIMENT and CONDITION POWDERS, and I consider them to be the most valuable remedies, and they should be universally used by stockowners."

WILLIAM MARTIN,
Ledge Road,
Newport, R. I.

"I have used your EMBROCATION and HOOF LINIMENT for several years now, and they both are exceedingly valuable applications for general use in many diseases of the limbs and feet of horses."

JAMES COLBROUGH,
Baltimore, Md.

"Please send me another dozen HOOF LINIMENT soon, as I am selling it right along since the carriage teams are returning home and they want no other kind after using yours."

W. W. ROBERTS,
1810 Market St.,
Philadelphia.